MILITARY APPLICATIONS OF SPACE
"THE INDIAN PERSPECTIVE"

MILITARY APPLICATIONS OF SPACE
THE INDIAN PERSPECTIVE

by

Gp Capt R K Singh

(Established 1870)

United Service Institution of India
New Delhi

Vij Books India Pvt Ltd
New Delhi (India)

Published by :

Vij Books India Pvt Ltd
(Publishers, Distributors & Importers)
2/19, Ansari Road
Delhi – 110 002
Phones: 91-11-43596460, 91-11-47340674
Fax: 91-11-47340674
e-mail: vijbooks@rediffmail.com

Copyright © 2014, United Service Institution of India, New Delhi

First Published : 2014
ISBN: 978-93-82652-26-7

CONTENTS

FOREWORD

I compliment **United Service Institution of India** to identify, commission and support a research project on Military Applications of Space. The topic is very important and assumes high relevance to the present times keeping in view the tremendous advancements in space technology achieved by some developed countries and in particular by our immediate neighbour – China and their intended use of such technologies for their national defence. The topic assumes further relevance when advanced space technology has been used by these countries for military applications to enhance their overall strategy of military deterrence and offence. This new dimension has further complicated our security scenario by pushing it beyond the conventional and traditional military balance. Over the last couple of years, India has made tremendous progress in space technology, thereby building our own space assets in the form of various satellites in LEO, MEO, and GEO orbits for applications in communications, broadcasting, surveillance, earth resource mapping, weather forecasting, navigation etc. including our own satellite launch vehicle technologies. India has also simultaneously developed matching ground space assets. With these developments and resulting assets, both in space and ground, India is increasingly dependent on these assets for its strategic and economic activities. It is, therefore, imperative to protect these space assets especially from strategic and security point of view by developing both defensive and offensive capabilities. This leads to militarisation of space, which may be a necessity rather than a choice for us. We need to develop such space technologies especially when our neighbour has explicitly demonstrated such capabilities. Such defensive and offensive space technologies should not only be state of art but also test proven. These, however, will depend on suitable national policy guidelines, which need be formulated by the Government at the earliest.

Group Captain R. K. Singh in his present book titled "Military Applications of Space-The Indian Perspective" has very ably dealt with the subject. He starts with introducing the subject, mentioning space as a strategic domain, historical perspective of military application of space, and various offensive and defensive capabilities. He, then, elaborates China's progress in space capabilities, and further discusses options for India in space security in the present times. He gives various options of offensive and defensive capabilities for appropriate policy formulations for India to secure our space and gives certain recommendations for the same. Group Captain R. K. Singh has extensively studied the available literature and has developed this comprehensive research work in the form of the present thesis. I am sure that this thesis will be used as excellent reference by other researchers.

Dr. V. K. Singh

Former Scientific Advisor, NTRO

office of the National Security Advisor

PREFACE

The world is moving at a very fast pace in every realm and the space is no different. Ever since the search of mankind for "the ultimate high ground" has crossed the boundary of earth's atmosphere and the first man made satellite 'Sputnik" was launched, this new domain started getting militarized. The military applications of space brought in a RMA in 21 st century, which every space player started to capitalise towards its military use. Not satisfied by their quest to have an edge over the adversary, the space faring nations have now moved ahead from militarization of space to its weaponisation. China proved its point by demonstrating its ASAT capability in 2007.

This book examines the paradigm shift brought in the strategic domain of space by testing of ASAT by China. The economic utility of space based systems has compelled the space faring nations, including India to secure their space assets. The proposed "Indian Space Security Architecture" has been discussed at length and recommendations have suitably been made for developing offensive and defensive counter space capabilities to meet the emanating security challenges to Indian space assets from adversaries.

The objective of the book has been to highlight the military applications of space, with special reference to China, and to recommend ways and means for India to secure its space assets by developing capabilities and technologies suitable; credible and potent enough to deter the adversary from causing any threat or damage to the space assets.

ACKNOWLEDGEMENT

'Military Applications of Space' was written as a Research Paper based on my passion for the subject. I am indeed convinced that the 'Space, the Ultimate High Ground' ' is going to be the new arena for future military conflicts and that the space assets will play a pivotal role in future warfare. It, therefore, becomes national imperative to strengthen militarily capabilities in space, so that India's defence preparedness is maintained to secure its national interest. The research on this contemporary subject has been very challenging, yet interesting. The militarization of space encompasses a whole range of military activities like communication, navigation, surveillance, monitoring, remote sensing and earth observation, each one of it in itself is a vast subject to cover. Therefore, an endeavour has been made to highlight the importance of each facet of militarization of space and dovetail it into the strategic realm and thereafter analyse the contemporary developments in each field and arrive at suggesting recommendations for utilization of space by India to create an **"Indian Space Security Architecture."** This security architecture in space will be able to counter the plethora of security issues and emanating threats from across the Eastern and Western borders of the country.

I am thankful to United Service Institution of India for having offered me the opportunity to work on this contemporary subject, 'military applications of space', which is being given highest priority world over to seek economic and military utilities of space towards economic development and national security. Having accepted my research proposal, it was the painstaking efforts of Maj Gen (Retd) YK Gera, Consultant and Head (Research) and his regular motivation and unflinching support that eventually enabled me to complete my research work. I will always remain indebted to him for standing by me, anytime and every time, whenever I needed support and for providing necessary inputs to refine my project by his timely advice. It was his conviction and faith in me, that kept me motivated throughout my stay at USI and was instrumental in completion of my thesis. I also thank my fellow researchers at the Institution for their prompt and timely support for making my research meaningful. Here, I would specifically like to thank Gp Capt AK Agarwal, a fellow researcher for his moral support, technical expertise and computer related support towards my research project. He has qualitatively contributed in shaping my project through his valuable suggestions.

I would also like to sincerely thank Lieutenant General (Retd) PK Singh, PVSM, AVSM, Director, USI for associating himself with my project study and re-igniting my imagination on the subject. The brief, but close interaction I had with him on the subject during our visit to Japan Institute for International Affairs, Tokyo, not only added value to my presentation on the subject there, but has also provided few important inputs for my research work.

Lastly, I profoundly thank my Project Guide, Dr VK Singh for helping me to lay a solid foundation for my research work, which helped me to conceive and chapterise the project and thereafter for guiding me at every stage. His in-depth knowledge of the subject coupled with his unbiased and critical comments have helped in shaping my study and bringing interest in the work.

Apart from the academic and professional help and support rendered by the colleagues and specialists, It will not be fair on my part if I reserve my appreciation for the sustained homely support from my wife, Seema Singh and the all important encouragement and computer support by my son, Apoorv Parihar, who through their love and appreciation always kept me focused on my project. But for the computer skills of my son, I had almost lost all my research data due to crashing of my laptop at final stages, which was retrieved to a great extent by him, or else I would not have been able to submit the thesis in time. The ensuing interactions with my son during the crisis period has infused a new meaning to our relationship, which is another valuable gain I will adore and admire all throughout my life, that is indirectly attributable to this research work.

CHAPTER – I

INTRODUCTION

Innovative technology is a boon to mankind and will ensure his supremacy in space. History is replete with examples as to how people or nations have utilized the latest technologies to effectively exert their will on their adversaries and showcase their might and power. History is also replete with examples of how innovation exploited even by primitive people helped them to petrify their opponents by using pointed stones tied to a stick and employed as a crude spear/lance. Since then, as mankind progressed through the ages, innovation has guided him from the use of primitive weapons to modern firearms of today. With the passing of each decade, there has been a thrust for developing newer technology to gain a cutting edge over the adversary in the conduct of war.

The effectiveness with which the newer technologies have been able to turn the tide of war in favour of the pioneer, made it almost mandatory for each nation state to invest time and resources towards R & D to retain the cutting edge. Be it the discovery of gun powder for use in canons or the design of tanks and armoured carriers and finally graduating to solid state missiles of this generation aided with GPS and laser guided munitions, the battlefield has evolved today into a very complex sphere. So to say, the impact of science and technology on the battlefield has been phenomenal and is getting more complex by every passing day. The SMART soldier equipped with every conceivable technology to help him in appreciating the battlefield linked to a C4ISR system and the use of decision making aids like lap tops and precision weapons, will transform the battlefield in days to come.

Similarly, the advent of Air Power in the year 1903 marked a new dawn for another innovation in the battlefield. The pioneering use of Air Power in the World War–I brought in all together a new concept of war fighting. The rapid development in the field of aviation technology from bi-planes to jet aircrafts and now to multi-role fifth generation aircrafts, have singularly changed the course of war at each stage. The standoff precision guided munitions (PGM) of today complimented with Cruise and ICBMs guided by inertial navigation systems have given the capable nations the cutting edge in modern wars. **Op Desert Strom** and **Op Iraqi Freedom** in Iraq, Op **Enduring Freedom** in Afghanistan, the effective use of Air Power in Kosovo and the very recent operation against Libya by the NATO Forces are examples of successful use of air and space based technologies for decisive outcome in war.

The next phase of the scientific development in the field of satellites and their launch mechanisms has once again revolutionized the war fighting capabilities of nations. Satellites have provided the elite nations with "eyes and ears" in space. Today, satellites can monitor the battlefield and provide latest pictures of less than 1 sq meter resolution. In addition, satellites provide navigational facilities with pinpoint accuracy (GPS) and seamless communication for better conduct of war.

The development of new technology is one thing, but the efficacy and the effectiveness with which they are incorporated into the war fighting capabilities of a nation is a remarkable innovation for military. The military use of space is expanding with every passing conflict as emerging technologies afford greater capability for manipulating the environment for pursuance of military activities. Up till now the space based assets were mainly aimed at "force enhancement" missions like observation, communication, navigation, meteorology, etc which allowed the terrestrial military forces to conduct their military affairs more efficiently. Thus most military space missions were auxiliary to other more direct military activities. For example the civilian earth observation satellites were also used for military remote sensing; civilian communication satellites have been known to carry military transponders.

However, as military/commercial reliance on satellites grows, so is the awareness that the space based assets have become the new centres of gravity for trade and commerce and also military activity. This has fuelled the quest for development of technologies for protecting one's assets in space as well as denying the adversary the use of space. With the development and testing of ASATs by China in 2007, there has been a paradigm shift in the fourth domain which perhaps may lead to a new Space Race. Commenting on this Michael Katz remarked *"This is a brutal effort in controlling the realm of space for one's own benefit, while denying it to the adversary"*.[1]

Militarisation of Space: An Insight

There is a strong belief that in the years to come, the world will witness another transformation in the conduct of war in the relatively new domain of space. It is this fourth domain which has generated immense interest post-Cold War, which the emerging power like China has focused on to seek asymmetric parity over the much stronger and technologically advanced superpower, the US. The emerging RMA in 21st century, which is significantly governed by space technologies, has already started shaping the way for drafting of new doctrines. This has resulted in militarization of space in order to gain strategic and tactical advantage over the adversaries.

Remote Sensing, Aerial Photography, Surveillance and Reconnaissance, Navigation, Broadcasting and Communication, and Scientific Experimentation are civilian uses of space technologies. However, the dual use nature of these technologies helps nation-states to exploit them for military purposes. Over the last few decades the advanced nations have successfully used space technologies towards military planning and even in military conflicts. The 1991 Gulf War, NATO intervention in Kosovo, the post 9/11 US involvement in Afghanistan and the 2003 US intervention of Iraq, they all have seen the military use of space technologies with amazing success. During these wars, the military use of space provided inputs for weather monitoring, communication, navigation and intelligence gathering. If one follows the series of operations world over, consequent to the First Iraqi War, there has been an exponential increase in space

1 Michael Katz in "The Final Frontier" p-13

technologies towards war fighting with remarkable success. Infact, with every successive war around the globe, the armed forces are getting over-dependent on satellites and the space technology.

In today's world, most requirements of military operations are driven by space technologies. The importance of Command, Control, Communication, Computers and Intelligence and Information, Surveillance and Reconnaissance (C4ISR) systems driven by space technologies has become indispensable to the military leadership. These C4ISR capabilities are being developed towards specific military applications in the fields of telecommunications, military information networking, electronic intelligence gathering, photographic reconnaissance etc.

The demonstration of US space capabilities during the 1991 Gulf War revolutionized the war fighting in 21st century. It clearly highlighted among other things, what can happen when a nation that does not enjoy the benefits of space exploitation wages a war against one that has it. The US enjoyed a virtual monopoly on space-based surveillance, communications, and navigation support during the entire spectrum of Gulf War. The US with its network of highly capable electro-optical and radar imaging satellites was able to determine exactly where to attack and which munitions to fire, while avoiding enemy troop concentrations, thereby reducing casualties. Similarly, during the Kosovo conflict, Afghanistan campaign and the 2003 invasion of Iraq the overall concept of the US operations was dependent on the information received from space-based systems.

Since the Gulf War in 1991, the world has seen the military usage of space technologies mainly by the US and allied forces. The US enjoyed unrivalled space superiority, which gave it an asymmetric advantage over their enemy. The overwhelming advantage that space technologies brought to the US forces in the 1991 Gulf War ensured the US forces a swift victory, as a result, this war went down in the annuls of history as the 'First Space War'.[2] Thereafter, in all conflicts which followed, the US forces extensively utilized space based technologies for providing navigational support to their weapon

---

2 "Not Ready for the 'First Space War,' What About the Second", a report by Steven J Bruger of the US Naval War College

delivery platforms and smart weapons such as JDAMS (Joint Direct Attack Munitions). All these uses of space technologies for war waging fall into the category of militarization of space. Thereafter, this decade has seen the beginning of weaponisation of space which was demonstrated by the Chinese in 2007 with their ASAT test.

In peacetime, the advance nations use their space assets for intelligence gathering surveillance and communication purposes. Thus the militarization of space which involves use of space assets for military purposes is not a new notion. Instead, what is new is that the capability to jam or destroy the adversary's operational space assets, or to weaponise space. There is a subtle difference between 'militarisation of space' and 'weaponisation of space'. Militarisation of space entails using various space assets for purposes of information gathering or helping the military to undertake land, air and sea battles or in simple words, tools to enhance the war fighting capabilities of a nation. It does not in any way has a destructive potential by itself. But, the weaponisation of space signifies getting into the act of destroying space assets of an adversary, by using ground based or space based weapons. Also, the arming of satellites with weapons that would be used against ground targets or co-orbital targets will also be construed as weaponisation of space. Besides, the weapons used to attack missiles traveling through space could also be termed as 'space weapons'.

The way militarization of space is progressing with concentrated efforts, the world is likely to see a new era of fighting wars in space. On 11 January 2007, China successfully carried out an anti-satellite (ASAT) test in which the type of weapon used for the kill was a Kinetic Kill Vehicle (KKV). This is a non-explosive weapon, which was fired into space with the help of a ballistic missile. It hit the satellite, fragmenting it on impact.[3] As a result of this ASAT test, the Chinese added more debris to space which has put other satellites in danger of collision with them. Here it may be noted that the space debris have brought in another dimension to the space weapon in the form of 'space mines', which can target the adversary satellites in a

---

3   Mastalir Lt Col Anthony "The U.S. Response to China's ASAT Test: An International Security Space Alliance for the Future, Anti-Satellite Capabilities and China's Space Weapons Strategy" published by Progressive Management (2012)

MILITARY APPLICATIONS OF SPACE: THE INDIAN PERSPECTIVE

known orbit. Thus this test has questioned the world's earlier belief that space would never become a battleground in the future.

In fact, the Chinese ASAT test was not the first time that such a test was carried out. In 1959 and 1968 the US and the erstwhile USSR had tested anti-satellite systems in their efforts towards Cold War supremacy. The late sixties was a period when 'weaponisation of space' was a much debated issue. However, the last ASAT test before this recent Chinese adventurism was carried out during the mid-eighties by the US. But very soon the consequences of weaponising space were appreciated by both the Superpowers, and eventually they realised that such tests would cause huge amounts of space debris which could harm their own satellites. As a result, an unwritten understanding was reached between the superpowers that they would not attempt to "conquer" this last bastion of warfare. But, the Chinese ASAT test indicates that this 'space reality' may change. Such tests would boost the desire of space powers to engage in one-upmanship, which will lead to a race for weaponising the space.

A detailed study of the historical perspective of militarization of space reveals that the Chinese ASAT test in 2007, alone cannot be held responsible for creating ripples in the global space architecture. Over the years, the US has always taken an entirely divergent stand on matters relating to space security. The Bush administration wanted to enhance this asymmetry by placing offensive and defensive weapons into outer space. In January 2001, a Space Commission led by Donald Rumsfeld, had recommended that the military should *"ensure that the President will have the option to deploy weapons in space"*. In fact, going through various space reports, Mr Rumsfeld expressed the opinion that if the US does not put its space security architecture in place, *"space could be the next Pearl Harbour for the US"*.[4] As was expected, after examining this report, President Bush withdrew in 2002, from the 30-year-old Antiballistic Missile Treaty (ABM) with Russia, which had banned the placement of weapons in space.

---

4   Jean-Michel Stoullig wrote in Space Daily "Rumsfeld Commission Warns Against Space Pearl Harbor" accessed through http://www.spacedaily.com/news/bmdo-01b.html

The May 2007 report of the International Security Advisory Board (ISAB) on US Space Policy elucidates the importance of militarization of space and the necessity of space superiority or space control. The report highlighted

> *"The United States considers its space capabilities vital to its national interest, and, accordingly, will take the actions necessary to protect and preserve its rights, capabilities, and freedom of action in space. This requires effective deterrence, defence, and, if necessary, denial of adversarial uses of space capabilities hostile to U.S. national interests. The Secretary of Defence is specifically directed to develop capabilities, plans and options to ensure U.S. freedom of action in space and to deny such freedom of action to adversaries when necessary. This requires robust capabilities for sustainable US space control."* [5]

All recent US policies related to space issues indicate that it believes in freedom of action in space, which it considers is important for its national interest and thus has rejected the proposal to ban space weapons. However they would support discussions on space under the aegis of UN, as is being done in the case of European Union sponsored 'Code of Conduct in Outer Space'.[6] However when it comes to their national interest, they will not enter into any negotiations on space weapons or a weapon free space. On the other hand, the Chinese demonstration of destroying a satellite should not be considered as a one-off event. Their interest in weaponisation of space has been known for some time. After three earlier unsuccessful attempts, the Chinese eventually were able to successfully conduct their ASAT test in January 2007. As a result, the Chinese have proved that they are now definitely focused towards creating asymmetric ways and means to counter US supremacy in the world. Though it was an avoidable test, but US hegemony on account of it being sole power in the world,

---

5 International Security Advisory Board (ISAB) provides the US Department of State with independent insight and advice on all aspects of arms control, disarmament, international security, and related aspects of public diplomacy

6 Lele Ajey in "Decoding the International *Code of Conduct* for Outer Space Activities" published by Pentagon Press, 2012

compelled the Chinese to seek asymmetric means to deter their stronger and powerful perceived adversary. Having realized the US over dependence on space technologies for both war-fighting and economic development, the Chinese endeavour towards developing ASAT capability is a calculated move to challenge US might. This indeed worked in favour of China, which was eventually able to create hysteria amongst the strategic community of the world, who consider China as a threat to world peace.

According to another report of 2001, China had also ground tested an advanced anti-satellite weapon called 'Parasitic Satellite'. It could be deployed on an experimental basis and enter the phase of space tests in the near future. This ASAT system can be used against many types of satellites in different orbits like communication satellites, navigational satellites, reconnaissance satellites and early warning satellites. According to a 'Space Daily' report, this nano-sized "parasitic satellite" is designed to be deployed and attached to the enemy's satellite.[7] There are three components to the ASAT "parasitic" satellites system: a carrier ("mother") satellite and launcher, and a ground control system. During conflict, commands are sent to this satellite to interfere or destroy the host satellite. The cost of building these satellites is 0.1 percent to 1 percent of any typical satellite, but the impact and the effect towards destroying an adversary satellite is phenomenal.

It was also reported by the media that in September 2006, Beijing had discreetly used lasers to "paint" US spy satellites with the aim of "blinding" their sensitive surveillance devices in order to prevent spy photography as they pass over China.[8] The Chinese aim was not to destroy the US satellites but to make them useless over Chinese territory. It has also been reported that the US military was so alarmed by the Chinese ASAT test that it revalidated its ASAT capability in 2008 by shooting down their own satellite USA 193 with their Aegis Missile system.[9]

---

7   Cheng Ho in Space Daily, a web based portal on space, "China Eyes Anti-Satellite System" accessed through http://www.spacedaily.com/news/china-01c.html

8   A report by Phillip C Saunders and Charles D Lutes for the US National Defence College "China's ASAT Test Motivations and Implications"

9   News report by Reuters in New Scientist web based portal on 21 Feb 2008, "US missile

Though a strong condemnation and reaction emanated against the Chinese ASAT test from nations across the globe, but there was very little that the global powers could do about the test. This was mainly due to the absence of a space treaty regime on 'space weapons'. For the last few years many players in the global space arena are trying to work out an international regime under the aegis of the United Nations to check weaponisation of space. Although an informal international understanding persists to desist from placing weapons into space, but no mechanism is available to punish the violators.

The United Nations in 1958, shortly after the launch of the first artificial satellite, started to work on space policies to ensure peaceful use of Outer Space and strengthen the idea of 'space as a global common'. In line with these efforts a Committee on the Peaceful Uses of Outer Space (COPUOS) was set up by the General Assembly in 1959. The mandate for the committee was to review the scope of international cooperation for peaceful uses of outer space. It was also expected to study the legal problems arising from the exploration of outer space. The COPUOS now has 67 member states and makes recommendations to the General Assembly on peaceful use of Outer Space from time to time.

The important disarmament agreement to provide the basic framework on international space law is the Outer Space Treaty, which entered into force in October 1967. This is the second of the so-called "non-armament" treaties (first being the Antarctic Treaty), which guarantees cooperation between states in all peaceful uses of outer space. Unfortunately, the treaty only prohibits the presence of nuclear weapons in space and it cannot therefore address the issue of weaponisation of space. Another important space treaty called the Moon Treaty came into being in the year 1979. This treaty declares that the moon (including all celestial bodies) should be used for the benefit of all states and the international community. It also expresses the desire to prevent the moon from becoming a source of international conflict. Unfortunately, the treaty has not been ratified by any nations engaged in manned space missions, so it is still a

hits spy satellite" accessed through http://www.newscientist.com/article/dn13359-us-missile-hits-spy-satellite.html#.Ubx099gfh0R non-starter.

The negotiations in space arena in various international forums have remained un-productive over the last few years. The Conference on Disarmament (CD) has not been able to agree on the formation of an Ad Hoc Committee since 1994 to negotiate a convention for the non-weaponisation of outer space. The prevention of an Arms Race in Outer Space (PAROS) initiative is also on the UN agenda since 1982. However, the US and Israel are unwilling to cooperate with the international community on the issue of PAROS. The US has even argued that the existing multilateral arms control regime is sufficient, and that there is no need to address a non-existent threat. But this view point has now changed with the testing of ASAT by China.

Apart from the hostile attitude adopted by countries like the US towards the establishment of any space treaty, the proposed regime also suffers from the problem of defining weapons in outer space. This is mainly because the definition of 'space weapon' is not well articulated as almost anything can be used as a weapon in space to obstruct satellites or to damage them. There would also be technical and financial constraints for verifying any deviations and irregularities, because of the complex problems involved in the verification of outer space activities.

As a fresh approach to prevent weaponisation of space and to some extent peaceful use of Outer Space, a new proposal in form of Code of Conduct (CoC) for Outer Space was proposed by the European Union in 2008. This Code was reviewed and resubmitted in 2012 to the UN for further deliberations by major space players. The CoC is now with the UN for deliberations and discussions by a "Group of Governmental Experts" (GGE). The Group reviewed the numerous proposals submitted by different states in recent years, for possible transparency and confidence building measures in outer space, broadly covering measures related to rules of conduct, measures aimed at expanding the transparency of outer space activities, measures aimed at expanding transparency of space programmes, and mechanisms aimed at resolving concerns. The final session of the Group will take place from July 2013 in New York.[10]

In fact, space analysts like Michael Krepon and Michael Heller have strongly recommended for an expeditious negotiation of a 'code of conduct' between space-faring nations, so as to prevent incidents and dangerous military activities in space. Also, global cooperation is possible in various other areas of space activities. The International Space Station (ISS) is one of the finest examples of such collaboration, where 16 countries have come together to undertake scientific experiments in outer space on a made-to-order platform. Similar collaborations are possible (in few cases they already exist) in areas like Navigation, Reusable Launch Vehicles (RLV), Space Commerce (Launch Business), exploring outer space to study the cosmos and use space assets over problematic border areas (like Kashmir or South China Sea) for strengthening confidence building measures (CBMs).

Thus going by the experience and implications of China's ASAT test, there is a need to use this incident as an opportunity to evolve long and short term space policies. There is a need to establish a strategic balance among the bigger nations, so as to break the monopoly on the utilisation of space by a selected few. It thus needs to be understood that while the peaceful uses of space and satellites are developing at a frantic pace, thereby facilitating global information and communication, the most advanced military powers are calculating how they can pursue war in this environment, by utilizing these very space technologies. Thus the challenges for major space player like the US, Russia or China is to continue exploiting space for 'defence' without weaponising it.

---

10 A report published by the UN Office of Disarmament Affairs, accessed through http://www.un.org/disarmament/HomePage/factsheet/wmd/Outer_Space_and_Disarmament_Fact_Sheet.pdf

CHAPTER II

SPACE – A STRATEGIC DOMAIN NECESSITATING MILITARISATION

'**Space**' as defined in Britannica Encyclopedia is, "*the boundless, three-dimensional extent in which objects and events occur and have relative position and direction.*" Physical space is often conceived in three linear dimensions, although modern physicists usually consider it with time, to be part of the boundless four-dimensional continuum known as "space-time". The concept of space is considered to be of fundamental importance to an understanding of the physical universe although disagreement continues between philosophers over whether it is itself an entity, a relationship between entities, or part of a conceptual framework.

However from a military point of view, space is considered anything beyond the earth's atmosphere or better known as "outer space". Thus as per the definition given in Webster dictionary, '**Outer space**' (often simply called **space**) is the space beyond the Earth's atmosphere. *Outer* space is usually defined from the point of view of the Earth and is used to distinguish it from airspace and terrestrial locations. The term can also be expanded to include the regions outside other celestial bodies in the Solar System (interplanetary space) or the regions outside the Solar System itself (interstellar space). On a universal scale, outer space constitutes the void that exists outside any celestial body.

A more acceptable definition of space was given by Theodore von Karman, a Hungarian-American physicist who calculated that above around 100 km in the atmosphere, the air becomes so thin that a craft must travel at greater than orbital speed to stay aloft. This is too thin for aeronautic purposes, and as such, activity above 100 km is demarcated as astronautic rather than aeronautic.[1] Orbits below the Karman line quickly degrade, approaching or slamming into the surface of the Earth. Above 100 km, a sustained orbit is possible, though orbital speed must be maintained to avoid degradation. When the Karman line was established, numerous scientists made the relevant calculations to determine where the line was, and when the results agreed on 100 km, they eagerly agreed on it as an official demarcation. The fact that 100 km is an easy-to-remember number helped it for its future designation as the Karman line. This definition is accepted by the Federation Aeronautique Internationale (FAI), which is an international standard setting and record-keeping body for aeronautics and astronautics.

Why Is Space Important?

During the Cold War, the space race represented not just national pride, but national security as well. In the 1960s, the US Vice-President Lyndon B. Johnson stated,

> *"Failure to master space means being second best in every aspect, in the crucial arena of our Cold War world. In the eyes of the world, first in space means first period; second in space is second in everything".* [2]

Today, space exploration has even wider connotations. The European Union has assessed the importance of space as follows:

> *"A command of space is key to success in the world of modern technology. The use of space has today penetrated all fields of economic, social and cultural management to a degree that makes space vitally important to the European Union. The ability to continue to develop and use space infrastructures autonomously and competitively, including collecting and using data, is clearly a key priority for Europe"* [3]

1   Frederic P Miller, Agnes F Vandome and John McBrewster in "Karman Line" published by International Book Marketing Service Ltd , 2009

2   David W. McFaddin, Lt Col, USAF, Can the U.S. Air Force Weaponize Space? (Research paper, Air War College, Maxwell AFB, AL, 1998).

3   EUROPA, Towards a European Space Policy, 2002, n.p.; on-line, Internet, 7 March 2002, available from http://europa.eu.int/comm/space/intro.

During the past forty years, space has moved from exploration to public and private exploitation; in other words, it has become a medium not that different than the land, sea or air. Gordon Adams, Director of Security Policy Studies at George Washington University, puts it this way:

> *"Space is no longer a frontier, used and occupied solely by governments. From an environment in which only governments operated, largely for exploration and military purposes, space has rapidly filed with assets used for intelligence and military operations to civilian communications, to observation and commerce. Today, more launches are dedicated to commercial purposes than to military ones."* [4]

The global space economy has seen a phenomenal growth to $304.31 billion in commercial revenue in 2012. A substantial growth of 6.7% was registered from the year 2011 which had a total of $285.33 billion. This growth was at a time when business across the globe was affected by global recession. The year 2012 also witnessed a steady rise in the commercial space activity, which included - space products and services and commercial infrastructure. The frantic pace of space activities and its commercial gains could be gauged from the fact that from 2007 through 2012, the total space commerce has grown by 37%. Similarly the commercial space products and services revenue has seen an increase of 6.5% over 2011 and the commercial infrastructure and support industries have increased by 11%. [5]

The rise in space commerce is attributable to the fact that for today's generation using space assets has become a routine event, much as television has over the past fifty years. When we turn on the TV, we simply expect the picture and sound to be there; no one speaks with awe about how the video and audio is available to us. Many people start their day by driving to work in an auto with a

---

4   Paper by Laurence Nardon, in Satellite Imagery Control: An American Dilemma, (March 2002) at the French Center on the United States (CFE), Paris, France, March 2002).

5   "The Space Report 2012, The Authoritative Guide to Global Space Activity" accessed through http://www.spacefoundation.org/media/press-releases/space-foundations-2013-report-reveals-67-percent-growth-global-space-economy

graphic display that depicts their present location; directs them across town following instructions to a predetermined destination; stop to top up petrol by using a credit card at the petrol pump; and draw money from an automated bank teller machine from their account that could be from a different bank in another part of the country. They will think nothing about the technological wizardry, but this set of transactions—location, directions, link to credit card and banking accounts—have all been made possible by instantaneous access to multiple man made satellite constellations. These and other satellite systems can provide navigation aids to civilian airliners, identification of underground water and mark the destruction of forests in addition to numerous other routine daily services, something taken for granted. The failure of a single satellite in May 1998 disabled 80 percent of the Pagers in the United States, as well as video feeds to cable and broadcast transmission, credit card authorization networks and corporate communication systems. If the Global Positioning System (GPS), a multi-satellite constellation originally designed for military navigational assistance were to experience a major failure, it would disrupt fire, ambulance and police operations around the world; cripple the global financial and banking system and could also threaten air traffic control.[6] Space, therefore whether we realize it or not, plays an increasingly important role in everyday life.

The evolution of space from a frontier to an operating environment with multiple users raised a numerous new concerns for policy makers' world over.[7] Recognizing the importance of space to U.S. national interests, the US Congress chartered a review of national security space activities. Released in May 2001, "The Report of the Commission to Assess United States National Security, Space Management and Organization," better known as the Space Commission Report, found that:

---

6 Peter L. Hays, Lt Col, USAF, United States Military Space: Into the Twenty-First Century, U.S. Air Force Institute for National Security Studies, U.S. Air Force Academy, INSS Occasional Paper 42.

7 John M. Logsdon, "Just Say Wait to Space Power," ISSUES in Science and Technology, on-line, Internet, Spring 2001, available from http://www.nap.edu/issues/17.3/p_logsdon.htm

"The security and economic well-being of the United States and its allies and friends depend on the nation's ability to operate successfully in space. To be able to contribute to peace and stability in a distinctly different but still dangerous and complex global environment, the U.S. needs to remain at the forefront in space, technologically and operationally, as we have in the air, on land and at sea. Specifically, the U.S. must have the capability to use space as an integral part of its ability to manage crises, deter conflicts and if deterrence fails, to prevail in conflict." [8]

The military has long understood the significance of space, which is recognized as the ultimate "high ground" for military operations. Space provides the opportunity for surveillance without the issues of over flight, and instantaneous communications capability that enables Command and Control of forces across the globe. Secretary of the US Air Force, Dr James G Roche stated that,

"Space capabilities in today's world are no longer nice-to-have, they've become indispensable at the strategic, operational and tactical levels of war." [9]

Peter B Teets, the Undersecretary of the Air Force and Director of the National Reconnaissance Office and a senior Department of Defence (DoD) space official, emphasized the critical nature that space plays today when he remarked, *"I think the recent military conflict in Afghanistan has shown us, without a doubt, how important the use of space is to national security and military operations."* [10] Citing the crucial support rendered by space based assets during the Gulf War, General Ralph E. Eberhart, Commander-in-Chief, United States Space Command, pointed out that,

---

8 Paper by Laurence Nardon, "Satellite Imagery Control: An American Dilemma", at the French Center on the United States (CFE), Paris, France, March 2002.

9 "The Report of the Commission to Assess United States National Security, Space Management and Organization,"

10 Scott Elliott, TSgt, USAF, "SECAF: Space forces have become indispensable," Air Force News Link, on-line, Internet, 24 September 2002, available from http://www.af.mil/news/Sep2002/92402411.shtml.

"Most anyone involved in military operations, whether military or civilian, would tell you, space is becoming increasingly important. Looking back to how we leveraged our space assets in Desert Storm, compare that to Kosovo–or how we can leverage them even today as we have made advancements since Kosovo–and I think it is obvious how important and how much we rely on capabilities that are resident in our information that moves through space.".[11]

Or as General Lance W. Lord, Commander, Air Force Space Command, succinctly put it, *"If you're not in space, you're not in the race."* [12]

History of Militarisation of Space

The military use of space is not a new phenomenon. In fact, the access and utilisation of space has become a subject of national interest. In addition to the economic potential of commercial exploitation of space and celestial bodies, space is the 'ultimate military high ground'. Historically, space-based military assets have invariably been passive, concentrating on activities such as reconnaissance, communications, and navigation. Indeed, expenditure on space for military application of space has always been much higher than the civil spending. So much so that even some of the scientific exploration missions have been dominated by military objectives, such as the race for technological supremacy during the Cold War which led both to the first satellite (Sputnik, 1957) and first human (Yuri Gagarin, 1961) in space and culminated in the manned lunar programme (Apollo, 1963-72).

It may be appreciated that during the Cold War, hardly any offensive space-based weapon had been deployed. The closest it came was during the parallel anti-satellite (ASAT) programmes developed by the US and Soviet Union which started in the sixties. As a result, programmes for a variety of 'kinetic kill' vehicles were initiated including initiatives for ground-based laser systems.

---

11 Ibid

12 Gerry J. Gilmore, "Space must be top national priority, says SPACECOM chief," Air Force News Link, on-line, Internet, 18 September 2002, http://www.af.mil/news/Apr20010406_0480.shtml.

Specifically these included programmes such as nuclear pumped X-ray lasers, space-based optical lasers, radiation-belt weapons, ground-based reflected laser systems, and space-based interceptors. While many of these projects were not completed, the technology base which matured as a result, led to development and near term deployment of space weapons. However the spin off was that many of the main components of space-weapon systems found their utility for the civilian space sector. For example, telemetry, tracking and control systems for a remote sensing communications satellite are very similar to systems within a space weapon. Further, there were certain self-imposed restrictions such as the reluctance of the US to use kinetic kill ASATs that tend to create large clouds of space debris.

As developments in space started to show immense potential for space commerce, the focus was more on exploiting the space based technologies towards commercial use rather than its military application. However the fact that space technology was of dual use, the military application was dovetailed with its commercial use as it never hindered military operations. Meanwhile, the international community repeatedly reinforced its support for using space for peaceful purposes only. This position was codified early in the space age by the 1967 Outer Space Treaty (OST), through which 96 states, including the US and former USSR recognised the common interest of mankind in the exploration and use of outer space for 'peaceful purposes'. The OST explicitly prohibits placing weapons of mass destruction in space or weapons of any kind on celestial bodies. Further, in 2001, the UN General Assembly approved the basis for a treaty, establishing a permanent prohibition on space-based weapons by 156-0 vote (Resolution 56/535). In 2008, a joint working paper on preventing space weapons (PPWT) was introduced by China and Russia in the UN Conference on Disarmament (UNCD). The two countries proposed to conclude a new international legal instrument through negotiation to avert weaponization of space and a subsequent arms race in outer space, so as to preserve a peaceful and tranquillity of outer space.[13]

---

13 Report of Foreign Ministry of PRC, "China and Russia jointly submitted the draft Treaty on PPWT to the Conference on Disarmament" accessed through http://www.fmprc.gov.cn/eng/wjb/zzjg/jks/jkxw/t408634.htm

However, the emergence of China's fast track development and consolidation of space technologies focused towards military application in addition to its civil use, saw a renewed interest in weaponisation of space. With a background of inactivity and caution of the UN and its inability to conclude a treaty for the peaceful utilization of space, the emerging space players have in recent years begun advocating developing new space weapons with strike capabilities. The very essence of this thinking was driven by the urge to gain asymmetric advantage over a superior and more powerful adversary. In April, 2003, for example, the US Congressman representing NASA's Florida base stated his support for weapons deployed in space:

"We must adopt a doctrine that states that we as a nation will vigorously pursue the ability to project power to, through and from space against any aggressor". He also noted that, *"It would be inappropriate to deny ourselves this advantage simply because of romantic notions of some that space is some type of sacred place".*

In fact he US military started advocating a strategy to include the deployment of space weapons within a matter of a few years to maintain its military superiority in the 21st century so as to maintain the status of sole superpower. Accordingly, the military focus on utilization of space has been reaffirmed repeatedly in key US documents such as Air Force Vision 2020 and other related strategic planning documents.

Space : A Strategic Domain

Space 'the ultimate high ground', is another strategic domain. It offers an observation platform, a communication hub, it is host to a highly accurate positioning system, a medium through which ICBMs pass, a pristine scientific environment, and a vast untapped commercial frontier. With so much at stake in space, the competition and confrontation between space players is likely to be viewed strategically.

It is evidently clear that the military significance of space is inextricably linked to its value and utility for both civilian and military purposes. Whether we like it or not, military 'Principles of War' formulated with the experience of countless conflicts, are extending into the realm of space. The situation today is similar to the period of infancy of airpower, wherein, it was intensely debated by military leaders to segregate the new medium of airspace from the military and reserve it exclusively for a dedicated independent arm which evolved as Air Force in the last century. However the overall military significance is particularly important in structuring a stable status quo. For example, Antarctica is a military-free zone by international treaty, and a large part of it is unpopulated, hostile to life, and of unique scientific interest. If Antarctica offered greener pastures or could be strategically exploited like space, such a treaty may not have been signed.

The resources and other benefits that space has to offer could be a reason for any future conflict. It may be noted that in the present form of warfare, the military dominance on land relies on air superiority, and towards that the contributions from passive space-based systems in the form of battlefield intelligence, navigation, and communication are advantageous, but are not a necessity for victory. However, in the future, space dominance will invariably become a deciding factor as the capabilities of the ground forces will be more and more dependent on space-related systems, thereby leading to a tiered dominance with space at the top – 'The Ultimate High ground' (see figure I). [14]

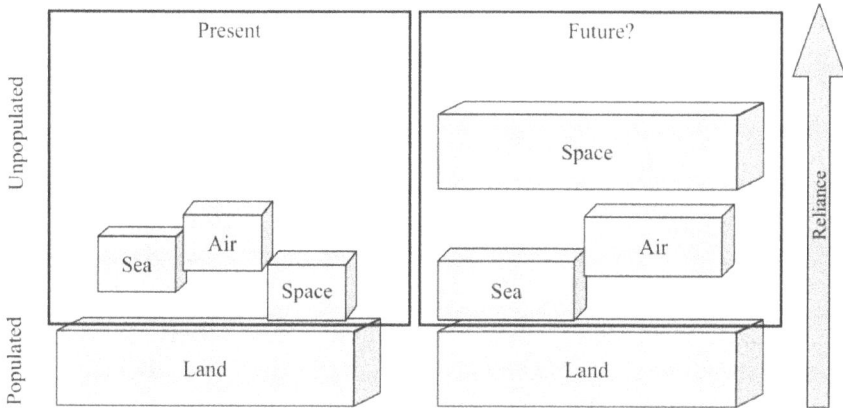

Figure I. Present and possible future reliance on different military domains

Space is an unpopulated medium; hence large-scale destruction in space will not imply loss of life that might otherwise occur on land. To achieve a military objective in any given conflict, the addition of air support to ground forces provides a 'sharper' tool to be used. By bombing selected targets the ground forces encounter lesser resistance from the adversary, in effect sparing lives that would otherwise be lost by ground assault alone. Similarly, the evolution of air systems to employ space-based GPS guidance has further developed this approach to gain an upper hand in the conflict situation. Today, a combination of air and space-based imaging can be used to identify vulnerable points, which can then be neutralised effectively and efficiently. This could perhaps defeat an opponent by preventing him from fighting. An opponent seeking to win by pure numbers in any future conflict may wish to begin by disabling current passive space-based systems. The risk of militarising space to protect this capability therefore opposes the risk of fighting bloodier wars.

The peculiarities of the Earth-Space boundary, the orbital mechanics and many other aspects of space determine the way space is used for defence, offence, and is itself defended by space faring nations. Unlike policy, force structure, and military technology, the attributes of space do not change. In fact, the activities in space have revolutionized the time scale factor in which things happen. The timescale of space transit is minutes, as compared to the other conventional arenas shown in Table 1.

Table 1 – The Transit-time Scale of Travel

Medium	Transit Time Scale	Probability of Defence
Land	Days	High
Sea	Days	High
Air	Hours	High
Space	Minutes - Hours	Low
Cyber Space	Seconds	Very Low

A satellite launch to a low earth orbit (LEO) takes 3-10 minutes, and will be the same for the fastest LEO trajectory time scale for an exchange of space-transiting weapons, such as intercontinental ballistic missiles (ICBMs). Further, the other categories of space-based weapons, like lasers, may occupy a more distant medium earth orbit (MEO), increasing the intercept time to hours and allowing the possessor a near instant strike from a weapon that has up to a third of the world's surface in its field of view at any one time. If we compare this time scale with the conventional attack of a similar magnitude, even the logistical build-up for such a major conventional military action will take months and the conflict per se, from days to weeks. This would be long enough for the international organizations like UN or a group of countries to intervene, and utilize diplomatic pressure to diffuse the situation.

Characteristics of Space Weapons

Space weapons started to appear, though in disguise, during the period of the Cold War. The US and the erstwhile USSR ventured to discreetly develop space weapon for supremacy in space. However, the threats to their own satellites on account of resultant 'space debris' compelled these superpowers to shelve their space programmes. However, the Chinese ASAT test in 2007 compelled the space faring nations to work on a treaty to ban deploying weapons in space and work for 'weapon free space'. However, there was no classic definition for 'space weapon' because it suited the Cold War adversaries to shroud their weaponisation programme in ambiguity. Even today, there has been no established definition for a 'space weapon'. Since even a small object if put on a collision path of the satellite in space can degrade or destroy it, hence it falls in the category of a 'space weapon'. With this definition even few kilograms of sand could also perhaps be used in space as a space weapon.

As space weapons are yet to evolve, there is intense debate within EU, Asian countries with space capabilities, Russia and the US to formulate an acceptable definition of 'space weapon'. The new definition is likely to include both weapons and targets located in space, direct and indirect applications of force, and temporary impairment as well as permanent destruction of any space asset located in space or on ground. A schematic diagram at Table 2 depicts the military activities in space, which are grouped into three categories and are characterized by generally agreed areas (black and white) as well as the grey areas. The activities in the white area are military applications of space that do not entail force application from assets stationed in space. The black area comprises technologies that fit the traditional definition of space weapons. The weapons in the interstitial grey area are more difficult to categorically classify because they constitute technologies overlapping both military and peaceful uses. These technologies are basically the dual – use technology which bring in the ambiguity factor and thus blur the line between space-based and space-transiting weapons. This ambiguity can be hypothetically exemplified through the use of temporarily-emplaced weapons that orbit for days to weeks, hence can be considered as a normal satellite with a normal payload which could also be tactically utilized as a 'co-orbital ASAT' as and when the situation demands.

Table – 2 [15]

SPACE WEAPONS (HISTORICALLY OR GENERALLY NOT ALLOWED)	INTERMEDIATE SYSTEMS	MILITARY ACTIVITIES NOT INVOLVING SPACE WEAPONS (GENERALLY ALLOWED)
[Key Words: **Degrade, Destroy**] - WMD or radiological weapons - Space -based directed energy weapons - Space based kinetic weapons - Anti-satellite satellites (ASAT) destruction or degrade other satellites	[Key words: **Deny, Disrupt**] - ASAT – Deny access to satellite or ground system, passive measures, encryption - ASAT – Temporarily interfere with satellite or ground system (cyber-attacks etc.) operation - ASAT – Disrupt operations of spac or ground segments permanently - Ground based directed (at space) wepaons - Nuclear wepaons for NEO defence - Ground based jamming - Suborbital intercept missiles for missile defence	- Communication - Navigation - Reconnaissance (Space based or high altitude platforms) - Space monitoring networks - Early Warning systems ICBM with suborbital trajectory - Suborbital delivery of troops or equipment

---

15 "Space weapons : the urgent debate " by William Marshall, et all at *Belfer Centre for Science and International Affairs, Kennedy School, Harvard University.

A representative example of this conception can be taken from a 1998 working group of the United Nations Institute for Disarmament Research (UNIDIR), which states:

"A space weapon is a device stationed in outer space (including the moon and other celestial bodies) or in the Earth's environment designed to destroy, damage or otherwise interfere with the normal functioning of an object or being in outer space, or being in the earth environment".

It may be noted that the activities in white are frequently employed in today's world by many nations (irrespective whether space faring or not). Some of the grey capabilities are maintained by a significant number of nations who possess space technologies. However, the systems within the black area are not fully developed or deployed, but have been the subject of intense national and international discussion due to their capability to create instability and destruction in on earth. These systems qualify to be appropriately defined as 'space weapons'. However, contrary to the traditional definition of space weapons, the technologies within the grey area are the ones that deserve immediate attention because they are the most likely assets which can be deployed in the short term. They will in turn exert the effects of other traditional space-based weapons. Hence there is a need for a distinctly clear line to be agreed upon amongst the space faring nations so as to suitably categorise all types of space weapon.

Today, objects in space can invariably be observed by nations possessing satellite tracking systems. The only limitation to the visibility and monitoring of the satellites is the limited stealth techniques which can still be used in space. Thus, an arms race of space based weapons could be disguised through 'dual use' technologies. It could occur on the ground in the form of space-transiting weapons that are discreetly stored until are required to be used. An apt example of this is the ICBMs of the cold war. Potentially, an outlawed and previously unknown space-transiting weapon need only be revealed by launching it, though one might reasonably expect to become aware of involved or widespread development efforts by means of intelligence gathering efforts or facility inspections.

The Economic and Security Implications of Militarisation of Space

Space has been militarized right from the advent of the space age. In an endeavour to reach the highest vantage point for conduct of war, mankind has eventually managed to reach space. Having reached the space, the militarisation of space was a matter of time. But the military applications of space also had economic spin offs. Thus there is a need that the space weapons must be assessed in light of contemporary economic and security developments in space technology. Disagreements over commercial, security and other peaceful uses of space may have important effects on the issue of space weapons. Let us take the example of Galileo, the European Union's embryonic satellite navigation system, which is in direct competition with the American GPS. The GPS data is used worldwide for anything from cellular telephones to Automated Teller Machines (ATMs) and a hoard of other services. With the United States experiencing trade (and now budget) deficits to the order of billions of dollars annually, the substantial profits generated by the GPS is an income that the US Government would protect and maintain in the future. Not only does the EU aim to capture a share of the GPS market, the Galileo system would also make the EU independent from US military data for modern warfare. As if this was not enough, the competition to the GPS system further grew because of consolidation of satellite based navigation (GLONASS) by Russia its former Cold War adversary and China the new entrant in space with its 'Beidou Navigation system'. The impact of this satellite based navigation system in commercial as well as military use has brought in competition between adversaries, but also between trading partners and allies. The twin drivers of economics and security which the satellite based navigation system has brought in the 21st century, has wider ramification in terms of economic rivalry leading to friction amongst nation states.

There is intense competition amongst the space faring nations over the Remote Sensing Surveillance satellites, another space application, which has a dual utility, both commercial and military. The proliferation of high resolution imaging capability has given a competitive advantage to smaller space players, matching the abilities and capacity of major players, thereby reducing the superpowers' exclusive hold on this strategic application of space. A positive outcome has been striking of a strategic balance and increased stability of global systems. However, during military operations these space capabilities may become a source of confrontation so as to gain tactical advantage over the evenly matched adversary. The 1991 Gulf War witnessed a clash of interests and a warning by the US to the French over sharing of data from the French commercial remote sensing satellite SPOT, as this data could help Iraqi forces in appreciating the locations of US forces and their strategic assets.

The Utility of Space Weapons

It is important to appreciate the strategic dimensions and the impact of deploying space weapons. It may lead to a counter -productive 'space race' between two adversaries with no one gaining tactically. Therefore, the choice should not be reduced to a question of whether the required technological capacity, financial wealth and political will is available, since outcomes of a war will not be dependent solely on the space weapon. Contrary to this, the result will emerge out of a strategic interaction between all the players. Whether a dominant state will enhance its advantage or gain national security by acquiring a new space weapons system therefore depends on how the other states are responding to the threat perceived to be emanating from space weapon.

In today's scenario, regardless of its power, a dominant actor cannot determine the outcome of war unilaterally. On the contrary, without due regard to the likely responses of other states, the decision and rational choice of a dominant actor to make the first move could result in a collective opposition from other states. This situation may lead to an impasse, without any significant advantage to the dominant actor. The US example in this context is worth quoting, wherein during the 2013 rhetoric by North Korea, which threatened not only South Korea and Japan, but also delivered a direct warning to the US after the US placed naval and space assets in Japan to counter the North Korean activities. However, in ensuing face off and Chinese collusion with North Korea, the US was not in a position to utilize its space assets to neutralize missile threats from North Korean.

Thus it would be prudent for any potential dominant actor to carefully consider the probable response and counter operations of other states in response to the placement of its weapons in space. In fact due to globalization, the responses adopted by the group of nations against the dominant player may be un-imaginable and un-manageable. In fact the responses may have a negative impact on global security as they might lead to the risk of starting an arms race with space weapons. The states should also consider the likelihood of spill-over effects into other strategic areas before venturing on to placement of weapon in space. The impact on nuclear strategy is particularly important to assess. Space weapons, along with comprehensive information warfare, may supplement nuclear deterrence as a strategic policy. This strategy could provide the post-nuclear deterrence paradigm for the nation states in 21st century. It may be noted out of experience of the prolonged 'Cold War' that the nuclear weapons have proved to be only a 'strategic weapon'. On the contrary, the space weapons will prove their worth as 'tactical weapons'. All said and done, the new generation of space weapons will reduce the overall reliance on nuclear weapons by the nuclear states. This is a positive impact, in view of the holocaust witness by the world in World War II. On the other hand, due to an increased military gap between the dominant state and other nations, the move could also lead to an increased likelihood of use of nuclear weapons by countries as a last resort and lower the threshold for using a nuclear weapon in a conflict.[16]

Proliferation of Space Technology

The lure of "Ultimate High Ground" motivated many nations to take up space programme for strategic as well as tactical gains over their adversary. The selective proliferation served the interest of the super powers during the cold war. However, with advancement of science, many nations took up space for military use and few like India and China took up space for societal good. However as the capabilities improve, the nascent space powers proved their technologies and started to compete, though in a limited way, with the super powers. After the major space players like the US, Russia, EU, China, Japan and India, there are new emerging players like Iran, South Korea

16 Ibid 27-30

and North Korea, who have just announced their arrival in space by launching their indigenous satellite through their indigenous launcher.

Thus, it is just a matter of time that the space players will graduate to placing weapons in space for tactical advantage. In response to that, there would be no shortage of potential actors that might respond to a first move by any state. While the US and Russia lead in capacity, the European Union, China, Japan and India, all have the requisite technical capabilities for at least certain space weapons systems. Thus given a first move by any state, the US is likely to act quickly to ensure tactical dominance in this domain. This will definitely be responded by Russia, which would also act to counter the initial deployment of space weapons. Since the Russian strength is dependent on its missile forces, any attempt to move from nuclear deterrence would reduce its power visa viz the US, unless it matches the US deployment of space weapons.

China, the other formidable space player will definitely like to balance the deployment of any space weapon by the US. Since China is investing heavily in space, it would like to seek parity in deployment of any space weapon which may threaten its space assets. It is unlikely that China will like to be restricted by any nation in space activities, unless there is a binding mechanism for all the space faring nations. It has proposed a treaty (PPWT) in the UN Conference on Disarmament for banning space weapons. Unless there is a regime for ban on space weapon, in all probabilities, if one strong player on the international arena gets too powerful, then the other smaller players may combine to produce a counterbalance. Thus there are possibilities of new alliances likely to be formed to counter a common adversary which is too powerful for individual nations. This is a distinct possibility in the Asian sub-continent where China's aggressive behaviour and posturing against its smaller neighbourhood, may result in Vietnam, Philippines, Brunei, South Korea and Malaysia joining together to form an alliance against China. In future, even India may join such an alliance, if the Indo-China issue of land border is not resolved and China resorts to forceful incursions and occupation of territory. Thus a dominant state should not only consider the chance of single nations countering their actions, but should also factor in the risk of many nations getting together to counter the perceived belligerent action of the dominant state.

CHAPTER - III

HISTORICAL PERSPECTIVE OF MILITARY APPLICATIONS OF SPACE

From the very beginning of the history of warfare, winning the high ground for military advantage has been one of the aims of every military campaign. Each side carefully plans its tactical moves to enable capture and retention of 'the high ground', which gives them a decisive edge leading to victory. In the earlier era of bow and arrows, fortifications were built on high points as a part of a strategy, with walls that permitted archers to rain down deadly volleys of arrows on the attacking enemy. On similar lines, mobile towers served as siege weapons. In fact ships of earlier days were equipped with crow's nests that enabled long-range reconnaissance. Hot air balloons were first lofted by Napoleon to observe troop movements. Aircraft were initially utilized for high level reconnaissance, which, with advancement in aviation technology was quickly followed by aerial battles.

The evolution of aircraft revolutionised warfare during the 19th century, and ushered in a new strategy of warfare appropriately called 'command of the air'. In fact, airplanes turned out to be a RMA of the 19th century. As aircraft evolved and became more powerful and sophisticated, it brought in a paradigm shift in war-fighting. Aircraft started going higher and higher for reconnaissance over enemy territory. However, the shooting down of the US high altitude reconnaissance aircraft, the Lockheed Martin U-2 Dragon Lady (which had a height ceiling of 70,000 ft)[1], by a Soviet SAM missile during the peak of the Cold War, led the quest for safer observation

---

1 Data accessed through http://www.militaryfactory.com/aircraft/detail.asp?aircraft_id=51

into the realm of space. This started the race for space between the Cold War adversaries, which was considered the "Ultimate High Ground". The Initial attempts towards control of space were led by both, the US and the Soviet Union, during which both conducted exercises for controlling the space with nuclear and conventional devices such as anti-satellite weapon (ASATs). Thus the 60's saw the beginning of militarisation of space by the two Super Powers, to gain advantage over the other. Thereafter the next phase was weaponisation of space with possibilities of positioning weapons in space for decisive military advantage. The Chinese ASAT test on 11 Feb 2007 saw a renewed effort towards weaponisation of space and the counter moves from the US followed in 2008 by testing Aegis Missile System to shoot down its own falling satellite USA 193. This transition from militarisation to weaponisation of space seems to be the next step in this endless struggle to gain higher ground. The idea of placing weapons in space can be found first in science fiction stories of 20th century, but it was not until World War II and the Cold War that such concepts became reality.

After the launch of the first satellite by erstwhile USSR, satellites became more sophisticated and complex and eventually evolved to become carriers of 'military payloads'. Substantial military activities were conducted by the Cold War adversaries in space through satellites. Space thus became an operating locale for many military satellites (such as imaging, navigation & communications satellites). With advancement in missile technology, long range ballistic missiles started to use space as a temporary transit medium for weapon delivery. However, no nation ventured towards permanent placement of operational weapons in space. This self-imposed restrictions were followed on account of fear that permanent placement of weapons in space (as opposed to non-weapon assets) would result in de-stabilization of the strategic balance between the Cold War adversaries, which eventually might lead to an arms race in the space. This would come at a huge cost and effort, that too without addressing the security scenario, inspite of the huge expenditure. Rather than making nations secure, developments in space miltarisation could never give them a sense of security. On the contrary, nations became more vulnerable to the militarized space technology. On the whole the entire gamut of militarization of space appears to benefit only the arms industry, which strongly advocates weaponization of space.

Evolution of Space : Its Relevance to Military Applications

For times immemorial, space has fascinated the human race. Numerous ancient myths and legends of different cultures recount incredible and at times visionary accounts of space being used for conflict resolution, apart from divine observation, intervention and punishment. Space-based military force application in terms of a celestial reprimand from the gods is a common recurrent theme in most of the ancient religious and mythical literature. Hence it is not surprising that ever since man started exploring this fourth dimension, he has been devising ways and means of utilizing it for military gains, which also had spinoffs in terms of economic, scientific and social advancement. Hence the evolution of space towards military application can be attributed to the pursuit of mankind towards better military capabilities like observation, surveillance as well as the improved efficiency of weapon delivery and attack on enemy positions. This pursuit was possible on account of holding the 'high ground', which in turn led mankind to go higher and higher, from the surface of the earth to air-breathing machines and then eventually into space by using rockets and satellites for enhancing their war fighting potential. Thus it is evident that ever since the dawn of civilisation, military doctrine places extraordinary emphasis on acquisition and retention of high ground for victory of military operations.

Early History of Military Applications of Space

In the early days, to attain high ground, humans fought for it for tactical gains over the adversary. Towards this, they developed crude propellants for launching fire pots, fire arrows, and the likes, which were a crude evolution of military ordnance for war. However, desire to reach farther and higher led to developments in the techniques and an improvement in the quality of propellants. By the 13th century, the Chinese became the pioneers in developing solid propellants like gunpowder, which revolutionised war fighting. The solid propellant improved delivery of fire-arrows and fire spears to much longer distances and gave it a respectable accuracy, which resulted in winning battles. Rocket technology continued to undergo refinements and development to a level that they soon reached "the ultimate high-ground," - space, by placing the first human made satellite 'Sputnik' in orbit on 04 Oct 1957. Thereafter, the space exploration evolved and it reached a stage where man could launch a satellite not only to the earth's orbit of different heights from LEO, MEO and HEO, but also a rocket to moon, to Mars and satellites in the sun and moon synchronous orbits.

Though the Chinese were the pioneers for early development of rocketry, it were the Indians who first inducted rocketry into their army and used them effectively in battles. In the battle of Panipat in 1761, the Marathas were reported to have launched barrage fires of up to 2000 rockets at a time against the Afghan forces. Rockets were also used predominantly in the battle of Srirangapatnam against the British forces by Tipu Sultan in 1792 and 1799. Tipu Sultan's army used rockets against the British to devastating effect. Consequent to the defeat of the British troops at the hands of Tipu's rocket contingents, the British studied these revolutionary weapons, refined them and used them under Colonel William. These Congrieve-design rockets were later used by the British against Napoleon's French armies. By the end of the 19th century, this rocket technology and its related military techniques formed the military strength of the European powers. It was in Europe that the rocket technology reached new heights and sophistications to evolve to the present levels of refinement, proficiency and efficiency.[2]

Going Higher And Farther

The European powers patronized rocket science and carried out lot of research for uses other than just for delivery of military ammunition. In 1906, a German named Alfred Maul successfully took aerial photographs of the ground by attaching cameras to a solid-fuelled rocket. Thus the innovations in the use of rockets led to tactical gains in battle. However, the arrival of aircrafts and refinement of artillery weaponry reduced the military utility of rockets. The use of aircraft in the war enabled acquisition of high ground with its related advantages. The early missions of aircraft were related to

---

2 Cliff Lethbridge in "History or Rocketry" Chapters 1-2

observation and delivery of military munitions. Later technological refinement carried out missions of communications, navigation, and weather observation. This brought in a new dimension to war fighting i.e., dominance in air. Thereafter, new theories and doctrines were written for control of the environment of air. This started the era of 'air warfare', which for time being dwarfed the developments in the field of rocketry. Apart from military contests for control of the air environment, the French introduced solid-fuelled La Prieur rockets by World War I. These solid fuelled rockets were designed to be fired from French or British bi-planes against German observation balloons. Eventually, the rocket technology with solid propellant proliferated in Europe, wherein the European powers effectively used them as weapons during the World War I. However, the defeat of Germany in the World War I lead to the Treaty of Versailles, which banned solid-fuelled rocket research in Germany.

World War II and the V-2 Rockets

In view of the ban imposed on Germany in terms of the Treaty of Versailles , after their defeat in World War I, Germany started experimenting with liquid-fuelled rockets in 1927, so as to circumvent the ban over research on solid propellant. By 1932 the liquid propellants started to demonstrate their potential for long-range artillery use. Eventually, a test vehicle designed and flown by Wernher Von Braun was accepted by the German Army for induction into their armed forces. In December 1934 Von Braun demonstrated another success with the test flight of the Aggregat 2, more famous as A2 rocket, which was a small model powered by ethanol and liquid oxygen. Many different liquid fuels were developed by the German scientists, but the use of ethanol as a rocket fuel was encouraged by Germany because of the shortage of crude-oil-based fuels.

The German designers had moved ahead in their research programs on A2 rockets and by 1936 they started working on both the A3 and A4 which saw huge improvements in the accuracy and range. For the A4 rockets, the Germans could ensure a range of about 175 km and an altitude of 80 km with a payload of about a ton. This increase in capability had come through a complete redesign of the engine by another famous German scientist, Walter Thiel. The first A4 flew in March 1942, flying about 1.6 km and crashing into the water. The second launch reached an altitude of 11 km before exploding. The third rocket launched on 03 October 1942, changed the course of history. It followed a perfect trajectory and landed 193 km away, and became the first man-made object to enter space. Since the A3 models were having problems, it was redesigned as A5 and test fired 70 times. The success of A5 saw its mass production which commenced in 1943 and was called, the wonder weapon *Vergeltungswaffe 2* (reprisal weapon 2), which became famous as the V-2, which created havoc over London during the thick of the World War II.[3]

The Cold War

During the Cold War, the world's two great superpowers, the Soviet Union and the US, spent large part of their GDP on developing space technologies with military applications. The aim was have the advantage of high ground. In 1957, the USSR launched the first artificial satellite, Sputnik-I. This ability to place satellites in orbit overlooking the desired part of earth stimulated intense research on the satellites and their launch vehicles. As a result, within a short span of time, different types of satellites were developed for different military applications. This intense space research in the United States and USSR, eventually led to militarisation of space, thereby starting the Space Race during the Cold War.

Both the Cold War adversaries developed their capabilities in space and by the end of the 1960s; they started deploying satellites for military applications. In fact the spy satellites were used by them to take accurate pictures of the rivals' military installations and nuclear assets. As time passed the resolution and accuracy of surveillance satellites improved and they became the preferred way of obtaining strategic/tactical value information. Alarmed with the threat potential of these spy satellites, both the US and the Soviet Union started working on anti-satellite weapons, so as to counter the capabilities of surveillance satellites by blinding or destroying them. Laser weapons, co-orbital satellites, as well as orbital nuclear explosion were researched with varying levels of success. Notwithstanding these ASAT tests, spy satellites found their utility

---

3   Kennedy Gregory P in "Germany's V 2 Rocket" by Schiffer Pub Ltd 2006

into monitoring the dismantling of nuclear tipped ballistic missiles in accordance with arms control treaties signed between the two superpowers. The use of spy satellites for monitoring of nuclear assets is often referred to in treaties as "national technical means of verification".

The superpowers developed ballistic missiles to enable them to use nuclear weapons across great distances. As rocket science developed, the range of missiles increased and intercontinental ballistic missiles (ICBM) were created, which could strike virtually any target on Earth in a timeframe measured in minutes rather than hours or days. In order to cover large distances ballistic missiles are usually launched into sub-orbital spaceflight. An intercontinental missile's altitude halfway through delivery is 1200 km, which is well into the realm of space. Thus space started finding utility in delivery of weapons over long distances and with accuracy and was accorded navigational support through satellites.

As a counter to the development and deployment of ballistic missiles, the Cold War adversaries started working on Anti-Missile programs. The 'Nike-Zeus Program' of the US in 1950 involved firing of Nike nuclear missiles against oncoming ICBMs, thus exploding nuclear warheads over the North Pole. This idea was soon shelved and work began on 'Project Defender' in the 1960's, which attempted to destroy Soviet ICBMs at launch itself, with satellite weapon systems orbiting over Russia. However this programme proved unfeasible on account of technological problems of that period and instead work began on the 'Sentinel Program' which used anti-ballistic missiles (ABM) to shoot down incoming ICBMs. However the major issue with the past ABM systems was that the interceptor missiles, though state of the art, required nuclear warheads to destroy incoming ICBMs. This shortcoming was addressed in the future ABMs, which became more accurate and utilized hit-to-kill or conventional warheads to knock down incoming warheads.

In 1983 American president Ronald Reagon proposed the "Strategic Defensive Initiative", a space-based system to protect the United States from attack by strategic nuclear missiles. The proponents of the "Star Wars" policy attribute the strategy of technology for hastening the Soviet Union's downfall. According to this viewpoint, Communist leaders were forced to either shift large amount of their GDP to counter the perceived "Star Wars" weapon systems or else helplessly watch their expensive nuclear stockpiles being rendered ineffective and obsolete.

The Soviet Union was also researching pioneering ways of gaining space supremacy. Two of their most notable efforts were the 'Fractional Orbital Bombardment System (FOBS)' and 'Polyus orbital weapons system'. The FOBS was a Soviet ICBM in the 1960s that once launched would go into a low Earth orbit where upon it would de-orbit for an attack on the ICBM. This system would create a path to North America over the South Pole, striking targets from the opposite direction from which NORAD early warning systems were oriented.

On 15 May 1987, the erstwhile USSR launched a heavy lift Energia rocket which carried a special payload, a prototype orbital weapons platform Polyus (also known as Polus). The final version of which according to some reports was to be armed with nuclear space mines and a defensive cannon. The Polyus weapons platform was designed to defend itself against anti-satellite weapons with a 'recoilless cannon'. It was also equipped with a sensor blinding laser to confuse approaching weapons and could launch test targets to validate the fire control system. However the Soviet attempt to place the satellite into orbit failed.

Post-Cold War Efforts towards Military Application of Space

The post-Cold War efforts towards militarisation of space was fixated around three main military applications of space technology which are

- Communication (C4) for Network Centric Warfare
- Intelligence (ISR)
- Navigation & Positioning (GPS)

The importance of communication and data transmission (of all kinds) amongst various stake holders in the military and the strategic realm, dictates availability of a secure and encrypted communication. The importance of communication is part of the emerging military doctrine of network-centric warfare, which relies heavily on the use of high speed and secure communications network, as this brings in connectivity amongst all soldiers in the battle zone and the military leadership in the command posts and the operational headquarters, with real time pictures of the battlefield. Real-time technology improves the situational awareness of commanders. For example, a soldier in the battle zone can access satellite imagery of enemy positions two blocks away, and if necessary e-mail the coordinates to a bomber or weapon platform hovering overhead while the commander, hundreds of miles away, watches as the events unfold on his monitor. It is the communication satellite which hold such a system together by creating an informational grid. An apt example of integration of communication network was **"Operation Geronimo"**, on May 2011, which culminated in the elimination of Osama Bin Laden in Abbottabad, Pakistan. Utilising the network centric capabilities based on space technologies available through a network of US satellites, the President of the US, his Secretary for State, CIA Director and staff connected with operations at the White House Situations Room, were able to observe the operation at Abbottabad in near real time.[4] This was made possible by space based communication technologies and its integration with the different elements of war fighting machinery of the army.

The second application is the technological advancements and developments of "spy" or reconnaissance satellites which began in the Cold War era. These Spy satellites perform a variety of missions such as high resolution photography (IMINT), communications interception (SIGINT), and covert communications (HUMINT). These tasks are performed on a regular basis both during peacetime and war operations. These satellites are also used by the nuclear states to provide early warning of missile launches, locate nuclear detonations, and detect preparations for clandestine or surprise nuclear tests. A nuclear test in 1978 by South Africa was picked up by a nuclear-detection satellite of the Vela type, which detected the clandestine nuclear detonation in the Indian Ocean.[5] This incident

---

4   http://www.cbsnews.com/8301-503543_162-20058792-503543/minute-by-minute-the-operation-to-get-bin-laden/

5   Daniel Alster in his paper "The Vela Incident: A Product of Political and Nuclear Cooperation

is famously called 'the Vela Incident'. In addition, the early-warning satellites can also be used to detect tactical missile launches. This capability was demonstrated during Desert Storm when America was able to provide advanced warning to Israel regarding the launch of Iraqi SCUD missiles.

The third major application of space militarisation currently in use is the Global Positioning System. The US military refers to it as NAVSTAR GPS - Navigation Signal Timing and Ranging Global Positioning System, the Russians call their system as GLONASS and the EU came up with GALILEO navigation and positioning system. The latest entrant to this elite club of nations, having their own navigation system is China with its 'Beidou Navigation System'. These satellite navigation systems are used for determining one's precise location and providing a highly accurate time reference almost anywhere on earth. The primary military purpose of the Navigation and Positioning System is to allow better command and control of forces through improved location awareness, and to facilitate accurate delivery of smart bombs, cruise missiles, or other munitions. The satellites also carry nuclear detonation detectors, like the ones in the US, which form a major portion of the United States Nuclear Detonation Detection System. This system is capable of providing a worldwide, highly survivable capability to detect, locate, and report any nuclear detonations in the earth's atmosphere, near space, or deep space in near real-time.[6]

China, the new space power in the making, began its Beidou Navigation Satellite System in 2000 with the sole aim of breaking its dependence on the US Global Positioning System (GPS) and creating its own global positioning system by 2020 for commercial and military use.[7] The Chinese Beidou system started its GPS services for Asia-Pacific in 2012 and plans to have global coverage by 2020. Inspired by the success of navigation system worldwide

---

Between Israel and South Africa"

6 A report from "National Security Space Road Map", accessed through http://www.fas. org/spp/military/program/nssrm/categories/nudetds.htm

7 A report by Chinese news agency Xinhua titled "China's satellite navigation system" accessed through http://www.gpsdaily.com/reports/Chinas_satellite_navigation_ system_live_Xinhua_999.html

and its relevance to the conduct of military operations, India felt the need of having an independent navigations system. Thus started a project called "Indian Regional Navigation Satellite System". This system IRNSS, which is still in the nascent stages, will be able to provide dedicated navigational and positioning services to Indian operators (both civilian and military), all across the Indian sub-continent and the Indian Ocean. As per an official release of ISRO,

> *"It is designed to provide position accuracy better than 10m over India and the region extending about 1500 kms around India. It will provide an accurate real time Position, Navigation and Time (PNT) services to users on a variety of platforms with 24x7 service availability under all weather conditions.*[8]

From the Fringes of Space to the Realm of Space

The A-4 rockets were eventually authorised for full-scale development in July 1943 by Hitler, which was later renamed *Vergeltungs/vaffeswi-2* (Venegance weapon-2) or simply V-2. These rockets used to pass through the lower fringes of space unchallenged at a speed of around 3,600 miles an hour before hitting their targets. The Allied forces had no means of intercepting them or to protect themselves. Thus the employment of space as a 'higher ground' provided a medium to deliver rockets on the target, thereby providing the military utility of space.[9]

The potential which the V-2 rockets demonstrated during the course of World War II motivated the Americans and the Russians to acquire the German rocket technology. Immediately after the war, both countries grabbed the opportunity to take control of the German Rocket Technology and lure their scientists and engineers to work for their own programmes. Under the 'Yalta Agreement' the Soviets were granted jurisdiction of Peenemunde, Mittelwerk plant area, Bleicherode, etc. where Von Braun's team and his V-2 facilities were located. However, the US out-raced the Soviets and vide **"Operation Paper Clip"** collected all the V-2s, V2 components,

---

8 Official Newsletter of ISRO of the period Jan-Jun 2012 accessed through http://www.isro.gov.in/newsletters/contents/spaceindia/jan2012-jun2012/article5.htm

9 Murray R Barber and Michael Keur in " Hitler's Rocket Soldiers" published by Tattered Flag Press, London.

documents and technical personnel, including Von Braun and his team and re-located them to the US shores. By the time the Soviets arrived, mostly unskilled lower echelon workers were available, who did not serve the Soviet purpose. This forced the Soviets to undertake their own research in development of their rocket technology.

Von Braun and his team launched the first V2 for the US on 16 April 1946, and eventually developed the US Redstone rocket which was intended to be used as a space launcher and as a tactical ballistic missile. Thus it can be said that the V-2s laid the foundation for both American and Soviet missiles. The first rudimentary Soviet ICBM-the *SS-1* Scud-was also nothing more than a refined copy of the German V2s.

On the other hand, the prevailing circumstances also dictated the need for both powers to stay abreast of each other's military activities. This was possible through violation of each other's airspace, which however was a violation of international law. This initially led to proposals like Eisenhower's "open skies" which was rejected by the Soviets, leaving the Cold War adversaries no choice but to go further higher in the sky and observe the adversary from an altitude high enough not to be shot down. This, hence lead to development of space- based observation surveillance and reconnaissance platforms. The focus of military application of space thus shifted beyond delivery of missiles to other applications like observation and communication. The US Air Force initiated 'Operation Feedback' in April 1951 to assess the possibility of using satellites for military observation and other purposes. By 1954, the operation was converted into military application through space, which included three aspects; reconnaissance via recoverable film systems (CORONA), infrared surveillance for missile launch detection (MIDAS)[10] and reconnaissance via electro-optical systems.

Unlike the US, the Soviet Union did not have a huge fleet of long range bombers, hence they took up developments in the field of ICBM and eventually produced relatively primitive atomic weapons that were bulky and required tremendous power to propel them across an intercontinental range. Thus came up the Soviet heavy-

10 Curtis Peebles, "Battle for Space", Beaufort Books Inc., Nevi York 1983, p. 52

lift launch vehicles which would allow delivery of ordnance over long distances. On 3 August 1957, the Soviets successfully flight-tested the world's first ICBM, the R-7 codenamed SS-6 Sapwood. [11] Based on their SS-6 ICBM booster, the Soviets launched the world's first artificial satellite, *Sputnik- I* (Traveller-1) on 4 October 1957, thereby heralding the dawn of the space age. With this launch, the superiority of Soviet military space technology was conclusively established. The Soviets conclusively demonstrated that they could now use the 'ultimate high ground' for delivery of munitions as well as for placing military platforms for observation, reconnaissance, etc. in space, the highest possible platform beyond the reach from earth, at least for that time period.

The impetus given to military uses of space systems beyond mere delivery of munitions continued un-abated. The period also saw the shooting down of U-2 , an American high-altitude strategic reconnaissance aircraft on 01 May 1960, while flying out of Pakistan over Soviet territory.[12] This incident and the capture of the pilot, Francis Gary Powers along with recovering the aircraft by the Soviets made the Americans suspend their U-2 programme. The proficiency shown by the Soviets in intercepting fighter aircraft and anti-aircraft missiles made aerial reconnaissance by the US an extremely risky proposition. The capability of carrying out reconnaissance from space which gave protection from being targeted by the Soviets, became a necessity for the US to develop. This resulted in the development of strategic satellites for reconnaissance, which were able to provide photos of Soviet missile installations from space.

The space race saw developments in space technology in both countries and they kept the tempo of super power rivalry by matching each other's satellite technology. On 28 Feb 1959, the US launched its first experimental reconnaissance satellite 'Corona' under the name 'Discoverer' which was followed by the world's first military navigation satellite 'Transite' on 30 April 1960. They also launched their first weather satellite TIROS- F in April 1960 and their first military communications satellite the 'Courier-lB'

---

11 Walter A. McDougalL "The Heavens and the Earth: A Political History of the Space Age," New York: Basic Books Inc

12 Ibid

on 04 Oct l960. The Soviets followed suit and in 1961 launched their first military reconnaissance satellite-Zenith-under the garb of Kosmos commercial satellite. Thus, the foundation for military uses of space was laid virtually immediately after it became possible to launch a space craft into a low earth orbit.

Utilisation of Space Technology in Conflicts

The early developments in the field of satellites took place to monitor nuclear missiles, their launches and the nuclear test. The primary role was to provide information about the launch of ballistic missiles, as the satellites were able to overcome the limitations of visibility on account of the Earth's curvature. Besides this, they were also able to reduce the delay in the warning time as compared to Earth-based radar systems. This led to perceptions on the strategic utility of space-based systems. As a result, within the half decade following the launch of first satellite Sputnik, the actual operational utility of the space based assets was demonstrated during the Cuban missile crisis, when the strategic employment of space-based observation was apparent. Although the US could not locate missiles in Cuba, yet the US satellite 'Discoverer' gave the US military and political decision makers reasonable estimates of Soviet nuclear delivery capabilities in terms of aircrafts and ICBMs. This conveyed to them that the capabilities of Soviet nuclear forces were quite limited, which in turn enabled Kennedy to call Khrushchev's bluff. Similarly, Soviet satellite Zenith provided data to Khrushchev that the US was actually positioning forces to attack Cuba and that the US Navy was moving into position to stop Soviet ships. The Soviets backed out, and the crisis was averted. This incident thus brought into focus the role of space systems for monitoring military activity, providing early warning to reduce the likelihood of surprise attack, and serving as National Technical Means of Verification (NTMV) to enable and enforce strategic arms control. Eventually this led to the development of the "Sanctuary" doctrine in the early years which was aimed at using space surveillance systems to make a nuclear war less likely.[13]

---

13 Lt Col Peter L. Hays, "Current and Future Military' Uses of Space," in a Conference on Outer Space and Global Security, 26-27 November 2002

The Asian Space Players

The space race between the two superpowers ended with the disintegration of the Soviet Union. The US was left with a large number of military assets, thereby providing it with worldwide dominance. Despite the new status of US in the world, there was no sole monopoly of space militarisation. The space activities took shape in the Asian continent, in which countries such as China, Japan, and India began their own space programmes. However, the European Union nations came together to collectively form the European Space Agency (ESA), which looked after their economic/ military utility of satellites and the business of satellite launch for their client states. They pooled in their resources to optimize the benefits of space exploration and thus worked to create satellite systems with military applications. One such system 'Galileo' which is the European equivalent of GPS navigation system, is for sure giving the economic competition to the US.

The Chinese started their space programme with the launch of their first indigenous satellite on 24 April 1970, thereby becoming the fifth country to send a DFH-1 scientific experimental satellite into orbit through an indigenous Long March rocket. On 26 Nov 1975, China launched its first recoverable satellite, which returned to earth three days later, and thus becomes the third country able to operate recoverable satellites. Having worked upon the benefits of satellite based technologies, China on 07 Sep 1988 launched a meteorological satellite FY-1A at Taiyuan launch base, north China's Shanxi Province and on 07 Apr 1990 sent its first AsiaSat-1 communication satellite into orbit through Long March CZ-3. Simultaneously they also worked upon the manned space project and on 20 Nov 1999 launched its Shenzhou (Divine Vessel) experimental spacecraft for the first time and the re-entry module landed safely in central Inner Mongolia Autonomous Region the next day. Thereafter they moved ahead with their project and perfected their space vehicles by successfully testing Shenzhou – 2 on 10 Jan 2001, Shenzhou – 3 on 25 Mar 2002 and Shenzhou – 4 on 30 Dec 2002, before becoming the third nation to send its first astronaut Lt Col Yang Liwei to orbit around the earth on 15 Oct 2003 in Shenzhou-5 spaceship through Long March 2F booster. Thereafter China achieved yet another milestone when Zhai Zhigang became the first Chinese national to walk in space on 27 September 2008.[14] Another distinction was achieved by China on 03 Nov 2011, when Chinese astronaut Liu Wang carried out the first docking of unmanned spacecraft Shenzhou-8 with space lab module Tiangong-1.[15] The success of the docking procedure makes China the third country in the world, after the United States and Russia, to master the technique, moving the country one step closer to establishing its own space station. Thereafter, China achieved its first-ever manually controlled space docking on 12 June 24 when its Shenzhou 9 piloted spacecraft and unmanned Tiangong-1 lab module connected together in the Earth orbit.[16] The docking moved the country a step forward in its plans to establish a space station within the decade.

Having appreciated the importance of space based technologies during the Gulf War I and the ever growing reliance of the US and Western forces on space based technologies, the Chinese leadership provided a fillip to their space activities. Besides the successful manned space endeavours, China worked its way to develop almost every range of satellites which have the potential for military application. They have launched satellites for communication, navigation, Earth Observation, weather observation and have ensured that there capability towards military use by PLA is retained. In order to gain asymmetric advantage over the much stronger and technologically advanced US forces, the Chinese carried out their first ASAT test on 11 Jan 2007, wherein they shot down one of their defunct satellite FY-01 by launching a KKE vehicle. With their developments in space, the Chinese have ensured that they can take on the US in the new domain of space in the 21st century.

The Indian efforts in space developments are equally laudable. Having started with the launch of first sounding rocket launch from

---

14 Hu Yinan in Jiuquan and Xin Dingding in Beijing in China Daily titled "Shenzhou VII launched for 1st spacewalk" accessed through http://www.chinadaily.com.cn/china/2008-09/26/content_7060750.html

15 News reported by Xinhua News Agency "titled "China accomplishes first space docking" accessed through http://news.xinhuanet.com/english2010/sci/2011-11/3/c_131226569.html on 04 Nov 2011

16 Space News from Collect Space accessed through http://www.collectspace.com/news/news-062412a.html on 24 Jun 2012

Thumba Equitorial Rocket Launch Station (TERLS), Trivandrum on 21 Nov 1963, India launched its first Satellite '**Aryabhata'** on 19 Apr 1975 through a Russian Cosmos-3M launch vehicle. From then on India has been indigenously building and launching a whole range of satellites for communication, remote sensing, weather and earth observation. The prime intent behind the Indian Space Programme was 'societal good and welfare'. However, having learnt from the US success in the Gulf War I and other operations in which the space based technologies were extensively utilized towards war fighting, India also started to re-orientate its civilian space programme towards the military application of space technology. The non-availability of adequate space technology for military application was felt by India the most during the Kargil War.[17] Thereafter, the available space technologies are being boosted to provide the armed forces some teeth in the modern war. Launch of RISAT satellites gave India a cutting edge 24X7, all weather surveillance capability.[18] Besides a Naval Satellite and India's very own Regional Navigation Satellite System are to start in 2013 and beyond, which will give independence to the Indian Armed Forces from the reliance on other powers for satellite based services.

Utility of Space : From Strategic to Tactical

As a result of the build-up of nuclear and conventional arsenal between the space powers and the reaction which it generated from non-space faring nations, concerns were raised on the peaceful use of space, which in turn led to the formulation of conventions and treaties. Thus, the antagonism of the big powers could never lead to a military confrontation of the World War II type. While both the powers avoided direct confrontation and mass destruction, the stress was to expand their spheres of influence with economic coercion or through covert regime change. They played through the gallery and believed in promoting local conflicts wherein their role was limited to that of indirect participants. Thus the world moved from an era of

---

17 Nair KK in "Space: The Frontiers of Modern Defence" published by Knowledge World 2006

18 Stephen Clark in "India Launches Surveillance Satellite in 'Grand Success' reported in news portal Spaceflight Now on 26 Apr 2012 accessed through http://www.space.com/15440-india-rocket-launch-surveillance-satellite.html

the super powers fighting wars directly under massive deployment and nuclear showdowns to localised conflicts under the indirect patronage of the super powers. In such a scenario the tactical utility of space-based assets towards ISR over the adversary became more and more relevant. There was an increased role for ISR satellites for monitoring military activities like movement of troops, armoured columns, assessment of the enemy's 'Order of Battle' and detection and tracking of multiple targets.

With the evolution of satellite technologies and capabilities, the role of observation satellites became more and more prominent, which was demonstrated time and again during numerous conflicts ranging from the Sino-Soviet border skirmishes over Damansky and Goldinsky Islands to the full-scale wars like the Indo-Pak war of 1971, the 1973 Arab-Israel War, the 1982 Falkland war and every major conflict till the most recent Libyan war. For example, during the Indo-Pak war of 1971, the Soviets launched high-resolution photo-reconnaissance satellites by KOSMOS 463 and 464 so as to enable observation of the areas on the eastern battlefield. This played a crucial role in effective decision making by the Indian Armed forces, thereby influencing the outcome of the war.

Similarly, during the Arab-Israel conflict of 1973, the USSR provided pictures of battlefield scenario with the help of six photo-reconnaissance satellites launched during the 21 days of hostilities. Three days prior to the outbreak of hostilities on 06 Oct 73, the USSR launched their first satellite for surveillance of the battlefield area. Within an hour of the Egyptian attack against the Israel across the Suez Canal, the Soviets had launched a high-resolution photo-reconnaissance satellite which recorded activities on the Egyptian front and also photographed the Syrian attack against Israeli positions on the Golan Heights. During the next three weeks the Soviets constantly monitored the situation through a series of satellites which they had strategically placed around the Middle East skies and provided the battlefield pictures and progress of war to the friendly Arab countries against the Israeli forces. It was through satellite photographs that the Soviets convinced President Sadat that the Israelis had made incursions across the Suez Canal and additional Soviet military aid was required to counter the situation. Later during the war when the Egyptian 3rd Army was cut off from Cairo and the situation continued to deteriorate, satellite photographic coverage confirmed the situation, and enabled the Soviets to press President Sadat to agree to a ceasefire.[8] Thus, the decisive role of observation satellites in conflict and war was established very clearly. Similarly the role or reconnaissance satellites in military conflicts can be appreciated from the fact that they constituted the largest category of military satellites in all the conflicts fought in the backdrop of satellite technology. To substantiate the fact, it is worth mentioning that the Soviet Union and Russia altogether have launched almost 780 photo-reconnaissance satellites till date.[19]

Military Applications of Space in Recent Conflicts

Operation Desert Storm

Following the invasion of Kuwait by Iraq in August 1990, the US initially launched 'Operation Desert Shield' to protect states like Saudi Arabia from further Iraqi aggression and followed it up with 'Operation Desert Storm' in January 1991. Though the US shield (of air, naval. land forces) was in place to protect the Saudi Arabia, the subject of interest is the space systems which were already in the orbit. They played a crucial role in crushing of the Iraqi forces by the US troops. Although the space assets played only a supporting role in the Allied build-up and actual combat which later followed, but what distinguished them from their role in previous conflicts was the fact that for the first time the entire range of space assets was employed in support of combat operations at all levels. The contribution of space assets for the victory of the coalition was considered so critical that the conflict was labelled the "First Space War".[20] The experiences and lessons of this war brought into the limelight the important role of space assets in decisively influencing the outcome of conflict. They were able to shape global perceptions based on the fact that for the first time experimental capabilities where demonstrated by the US satellites.

---

19 Brian Harvey, "Russia in Space: The Failed Fromier?" p 109, Praxis Publishing Ltd, Chichester~ UK, 2001

20 The US Air Chief Gen Merill McPeak was the first person to use this term

Desert Storm was the first war when the entire range of satellites for Intelligence Surveillance Reconnaissance (ISR), navigation, meteorology, communications, Early and Warning (EW) were available to the coalition forces simultaneously. It was the first occasion when the cumulative effect of over sixty satellites dedicated to the coalition effort was evident in influencing the outcome of war.[21] The support provided ranged from satellites providing a clear view of the battlefield to navigational satellites providing precise targeting, safe maneuvering, and precise munitions delivery information and so on. There were dedicated Met-Satellites providing weather inputs to support the entire spectrum of combat operations. They provided near real-time information critical to campaign planning on the ground, sea and air. Early Warning (EW) satellites like the Defence Support Program (DSP) basically designed to provide EW of Soviet intercontinental missiles were instrumental in detecting and shooting down incoming Scud missiles. Communication satellites provided reliable and near-total intra and inter-theatre communication links, enabling secure battlefield communications as well as communications between the Iraq-based coalition commanders and the US White House.

It was for the first time that the ever increasing demand for communication satellites during the course of Iraq War compelled the coalition forces to lease commercial space systems to cater for the increasing military demands. Coalition communication systems were reported to carry more than 700,000 telephone calls and 152,000 messages per day during the most intense part of 'Desert Storm'.[22] Out of this, satellite communications systems carried 85% of the total inter-and intra-theatre load.[23] Similarly, the demand for GPS systems also outstripped military demand.[24]

21 Sir Peter Anson BT and Dennis Cummings, "The First Space War: The Contribution of Satellites to the Gulf War, *RUSJ Journal,* Winter 1991, p 45

22 Lt Col Steven Bruger, "USAF: Not Ready for the First Space War-What about the Second" *Naval War College Review.* Winter 1995, p. 75

23 Ibid p 76

24 Lt Gen Thomas S Moorman USAF in "Space: A New Strategic Frontier" in Air Power Journal Spring 1992

One significant aspect which was noticed in this war is the fact that technology was clearly ahead of prevalent doctrines. Since the doctrinal concept for employing space systems had not been formulated by the coalition forces, it resulted in sub-optimal utilisation of space assets and their integration into operational use.[25]

Kosovo and Kargil

The two distinct wars in the era of satellites which need to be analysed regarding the employment space technology are Kosovo War and Kargil War fought in two different geo-locations. While the Kosovo war witnessed the largest number of space assets put in use to support the war effort of the coalition forces, in contrast the Kargil conflict was witness to not a single space asset being used, though at that point of time, India was an established space power. While Kargil was characterised by lack of information in all aspects ranging from intelligence on enemy locations, their numbers and weapons, weather inputs, etc Kosovo was characterised by a surfeit or space-based military information for the coalition forces.[26] Kosovo operations by the coalition forces is perhaps the best example for application of military power based on the inputs provided by the space technologies to conduct a swift and a decisive victory in battle.

Operation Iraqi Freedom

Space systems performed similar functions during 'Operation Iraqi Freedom' as they did during 'Operation Desert Storm', but with refinements in technologies and doctrines which were focused towards optimal exploitation of space assets. This allowed greater integration of space technologies with the conventional war-fighting capability. Unlike Desert Storm, space-based capabilities were integrated with actual military operations at all levels and the gains in combat effectiveness were apparent to the world. The success of 'Operation Iraqi Freedom' depended heavily on improved support and force enhancement capabilities provided by space-based assets. The whole intent was focused on bringing about an integration of the battlefield by placing air, space and supporting troops in the field as

---

25 Robert Kehler, "Space-Enabled Warfare," *RUSJ Journal,* August 2003, p. 68

26 Nair KK in "Space: The Frontiers of Modern Defence" published by Knowledge World 2006

well as other service components in the war zone.[27] The cumulative effect of the this integration is apparent when considering the fact that it was this combination of space as well as aircraft, precision weapons and updated intelligence which led to pin point attacks and precision bombings. It was the strength of space technology that allowed minimal co-lateral damage, such as an attack supported with inputs from space that lead to destroying only the fourth of seventh floors where the tactical target was located.

Space technology has been able to reduce the numerical demand of other military systems. For example, during 'Operation Iraqi Freedom', while the increase in the number of satellites was marginal, the coalition air forces employed only 1,900 aircraft, as compared to 2,500 aircraft in 'Operation Desert Storm'.[28] Though 542,000 coalition military personnel participated in Operation Desert Storm, only 350,000 personnel were used during Operation Iraqi Freedom. As compared to 10% precision-guided munitions in Operation Desert Storm, Operation Iraqi Freedom saw the use up to 68% precision guided weapons, leading to lesser wastage of munitions and reduced collateral damage.[29] This highlighted the utility of space assets towards war fighting in the 21 st century.

27 William B. Scott and Craig Covault "High Ground Over Iraq," *Aviation Week & Space Technology,* 9 June 2003, p. 45

28 Michael Knights, "Iraqi Freedom displays the transformation of US airpower," *Jane's Intelligence Review,* May 2003, p. 16

29 Ibid

CHAPTER IV

ENHANCING MILITARY CAPABILITIES THROUGH SPACE EXPLOITATION- OFFENSIVE AND DEFENSIVE SPACE CAPABILITIES

Space capabilities are becoming absolutely essential for national development, economic well-being, commerce, and daily life, besides becoming a crucial component of successful military operations. It is a well-known and accepted fact that space has emerged as an essential component in furthering a nation's comprehensive national power. India has a robust civil space programme which concentrates on scientific and development goals. As the space programme matures, employment of space for economic and developmental purposes is likely to increase, and as dependence on space assets and systems increases; the concurrent vulnerability of our country to hostile action seeking to destroy, degrade or deny our space capabilities so painstakingly built over the decades would increase. India's dependence on space for vital economic purposes has been growing rapidly, hence any serious damage or degradation would have a major negative impact on our national economic and security interests.

The lessons of history, on the other hand, are clear that wherever serious threats to national economic interests arise, military force would be necessary to protect them in the best manner possible. Military organisations have evolved as instruments of national power to protect national interests and investments. This generates the rationale for military involvement in space; besides the fact that space-enabled capabilities are the core of Revolution in Military Affairs (RMA) in 21st century which is aimed at enhancing terrestrial military capabilities and national defence. The kind of modern precision warfare witnessed during the First Gulf War is largely a by-product of this RMA which is aimed at combining the cumulative potential of air and space forces in terms of Intelligence, Surveillance, Reconnaissance, (ISR), communications, navigation, etc. for providing information dominance which is so vital to application of force, which in-turn enables decisive war-winning effects. [1]

Militarisation of Space : The Stark Reality

Space hasn't yet been weaponized but it is already highly militarized, thanks to a money-hungry arms industry and a commission started by Rumsfeld.[2]

~ Stan Cox , a space commentator

On 11 January 2007, a missile launched from China's Xichang Space Centre destroyed a satellite 537 miles above the Earth's surface. Although the target was a weather satellite belonging to China itself, the act clearly rattled the space and strategic community world over. One observer from a US think-tank rightly said, "the new space policy says we can defend the heavens with technology, but we can't, and the Chinese just proved it." It was perceived that the beginning of militarization had not only begun but had proliferated.

On 11 Jan 2001, precisely six years earlier to the Chinese ASAT test, the commission which was directed to assess the United States National Security Space Management and Organization, submitted its report to the US Congress. The Commission, which was headed by Donald Rumsfeld (who later became Defence Secretary to President George Bush's government), asserted that

"It's only a matter of time until there's all-out war in the heavens. We know from history that every medium -- air, land and sea -- has seen conflict. Reality indicates that space will be no different. Given this virtual certainty, the US must develop the means, both to deter and to defend against hostile acts in

---

1 'The Final Frontier" by Michael Katz-Hyman, Stimson Center in Arms Control Today, November 2004, p. 13.

2 Remarks by Stan Cox, a space commentator.

and from space -- and ensure continuing superiority". [3]

The apprehensions expressed by the Commission about the threat to US assets in space, eventually came true with the ASAT test by China in 2011. Highlighting the importance of space in modern warfare, and its role towards protecting the national interest, the Commission made some critical and time bound recommendations to the Congress. The gist can be summed up as below

> *"Military space officials will have to develop new doctrine and concepts for offensive and defensive space operations, power projection in, from, and through space, and other military uses of space."*

-- Rumsfeld Commission Report [4]

Military Application of Space

The military application of space expands with every passing conflict as emerging technologies afford greater exploitability of the environment for pursuance of military activities. Until the last conflict, the military applications of space were largely of "non-weapon" nature, which was focused in supporting the armed forces towards war fighting. Space-based assets were mainly aimed at 'force-enhancement' missions like observation, communications, navigation, meteorology, etc. which allowed terrestrial military forces to conduct military affairs more efficiently. Thus most military space missions were auxiliary to other more direct military activities. The capacity to deliberately cause damage to another party is not the main criterion for attributing a military character to satellites. Most present-day satellites (excluding ASATs) affording military capabilities or performing military functions are incapable of directly destroying or damaging another country's property. Apart from 'Early Warning' satellites which have a clear-cut military role, most of the other military activities can also be performed by civilian satellites and vice versa. For example, civilian earth-observation satellites are used for military remote sensing, civilian (even commercial) communication satellites have been known to

---

3 Report from the National Security Space Management and Organization of USA

4 Report by Donald Rumsfeld's Commission

carry military transponders, military navigation satellites have overwhelming civilian users, etc.

However, as military and commercial reliance on satellites has grown, so has the fact that space-based assets have become the centres of gravity which are likely to be targeted during war. This in turn has fuelled the quest for development of techniques for protecting one's assets in space as well as denying an adversary the use of space. Thus, while up to the last conflict involving space, space systems were mainly focused on force enhancement missions, the present focus has shifted to controlling the realm of space for one's own benefit while denying it to the adversary. The accent on military utilisation of space is gradually shifting beyond enhancement of military force capabilities to control of the environment and actual application of military force "in, from and through space".[5] The above trend is visible in the quest of space-superpowers like the US and China embarking on programmes aimed at space control and space projection.[6] Some of these include programmes like the Experimental Satellite Series (XSS) which seeks to use small satellites to manoeuvre around other satellites in order to inspect, service or attack. They also include Kinetic Energy Anti-Satellite (KEASAT) systems, Directed Energy: programmes as well as "Counter-Space" initiatives like the Counter Communications System (CCS) aimed at disrupting satellite-based communication used by an enemy for military purposes. The first of such CCS system was delivered to the US's 76th Space Control Squadron in the year 2004.[7] Apart from the above, a 'space based interceptor test bed' programme is also under way to develop and test space-based miniature missile defence interceptors. The Pentagon's Missile Defence Agency has already provisioned budgetary allocations for the same.[8] The concept broadly envisages a limited constellation of

---

5 Excerpts from "Report of the commission to assess United States National Security Space Management and Organisation", 11 January 2001, p. 16.

6 Michael Krepon, "Weapons in the Heavens: A Radical and Reckless Option", Arms Control Today, November 2004, p. 10

7 Jeffrey Lewis, "Programs to Watch", Arms Control Today, November 2004, p. 12. Also Nicole Gaudiano "USAF seeks weapons for counterspace capability" Defence News. 25 July 2005, p. 44.

8 GopalRatnam, "Killers from Space", Armed Forces Journal, June 2005, p. 24.

space-based interceptors of 50 to 100 satellites offering a thin boost/ascent defence against ICBMs and a multi shot mid-course defence against medium intercontinental range missiles. It is thus evident that space-based systems are presently in the process of transition from an era of militarisation to weaponisation.

Military Applications of Space by Nascent Space Powers

It needs to be borne in mind that the aforementioned transition is applicable only to nations like the US. Its closest rivals, the Russians and Chinese are yet to embark on any serious weaponisation programmes. This is mainly on account of the prohibitive costs and technological challenges. The Russians inherited the entire range of capabilities for force-enhancement missions from the former Soviet Union. However, since the 1990s, its capabilities have been severely degraded due to insufficient funding. As of 2004, Russia maintained military space programmes only in five areas of early warning, optical reconnaissance, communication, navigation, and signal intelligence.[9] With regard to ASATs, the former Soviet Union was the only country that developed and operationally deployed an anti-satellites system (ASAT), designed to attack satellites on low-earth orbits. However, the present Russian Federation is not known to have any operational ASAT systems.[10] As for the Chinese, though they are the undisputed leaders in Asia in relative terms visa-a-vis the US, their capabilities are nascent. [11] However, the world has taken cognizance of the successful launch of ASAT (a Kinetic Kill Vehicle) to shoot down one of its own ageing weather satellite FY-1C.

Other countries with known space-based force enhancement assets in operation include France (Helios image intelligence satellite and the Telecomm-2 communications satellite), Italy (Sicral communications satellite), Spain (Hispasat communications satellite), Britain (Skynet-4 communications satellites), Israel (Eros

---

9 PavelPodvig, "Russia and Military Uses of Space", Stanford University, Center for International Security and Cooperation

10 Ibid

11 The US Department of Defence, "Annual Report on the Military Power of the People's Republic of China" July 2003, p. 36 claimed that China was developing killer microsatellites based largely on a January 2001

and Ofeq imagery intelligence satellites), India RISAT and TES (photo-reconnaissance satellites) and Cartosat; Japan (commercial Superbird communications satellite system and Information Gathering Satellite); and South Korea (Kompsat-l remote sensing satellite). Thus apart from the US, Russia and China, most of the other II-tier space faring nations are yet to progress beyond rudimentary military space capabilities and force enhancement missions.

Space-Based Military Force Multiplier

The capabilities afforded by space systems, both economic and military, are towards force-enhancement and force multiplication, which eventually contributes to enhancing Comprehensive National Power. The military applications are as discussed below. [12]

- **Early Warning Satellites**. These are used to monitor enemy territory for military activity such as missile or satellite launches, missile tests as well as nuclear detonations. Space-based sensors are capable of detecting ballistic missiles almost immediately after launch and provide maximum warning time for retaliatory or counter action. They thus constitute a vital component of EW systems.

- **Observation/Intelligence Surveillance Reconnaissance (ISR)**. Their primary contribution is to enable 'space situational awareness' by providing information about a multitude of military activities by generating high resolution images of areas of interest, monitoring changes, strengths and locations of forces, etc. This includes the important sub-sets of Imagery Intelligence, Signals Intelligence, Electronic Intelligence, etc. Satellites fitted with requisite sensors are also used for ocean surveillance.

- **Communications.** These satellites enable military commanders to exercise command and control over their forces and to receive real-time information regarding the progress of a campaign or about possible enemy actions to a degree that were previously unknown. Apart from this

---

12 Rear Adm A.P. Revi, "An Integrated Strategic/Space Command Option", Indian Defence Review, Jan-Mar 2005, Vol. 20(1), p. 87-90.

space and terrestrial sensors involved in ISR, navigation, etc. generate enormous amounts of data. The transmission of this and other data for military purposes needs reliable and secure communication which is provided by communication satellites.

- **Navigation Satellites.** These are used to provide accurate targeting, positioning and navigational location information to users for strategic, operational and tactical requirements. These help military forces to precisely manoeuvre, synchronise actions, locate and attack targets as well as locate and recover stranded personnel and many other actions. They have profoundly improved military capability of reconnaissance, accuracy and safety of weapon delivery platforms as well as weapon delivery itself, in addition to accurate deployment/re-deployment of military forces.

- **Meteorological Satellite.** These satellites provide data on atmospheric water vapour, temperature and other weather phenomena. They are instrumental in determining the most appropriate moment for attack; they also monitor meteorological conditions during flight over target areas as well as providing real-time weather information over the area of interest. They provide data about cloud cover so that satellite reconnaissance missions can be planned efficiently.

- **Geodetic Satellites.** These produce maps of the earth by using photographic and radar techniques. They also provide data about the earth's gravitational and magnetic fields which enable trajectories of ballistic missiles to be predicted accurately and are essential for the guidance systems of cruise missiles.

Advantages and Spin-Offs of Space Technologies

Apart from these apparent capabilities, numerous other military advantages are obtained through the space based assets, which further increase the Comprehensive National Power of a space faring nations. Few of these military capabilities are discussed below :-

- **Battlefield Transparency.** Battle field transparency is enabled by a coherent mix of space, air and surface-based sensors complemented with multi-sensor data fusion capability to enhance 'situational awareness' at all levels. This capability is vital to combat decision making and force as well as weapon employment. [13]

- **Freedom of Operations.** This situational awareness in conjunction with the enormous information afforded by air and space platforms would provide vital inputs for successful prosecution of military operations. This in turn would ensure that own forces are used discriminately for delivering weight of effect at the right time and right place thereby enabling effective operational employment. This information would also expose friendly forces to lesser risks by navigating them safely on to the target. This freedom of action would create operational opportunities for own forces while conversely limiting the adversaries. [14]

- **Tactical Surveillance.** Space offers the potential for safe and continuous tactical surveillance over designated areas of interest. Apart from the vertical depth and strategic breadth of vision they would offer enormous intelligence inputs, which could ensure that the adversary is under constant surveillance thereby reducing his opportunities of surprise and seizing the initiative.

- **Deterrence.** Deterrence can be created by continuous surveillance of porous borders to prevent illegal cross-border movements and transaction. In fact, even during sensor switch-off periods, the psychological impact of this continuous multi sensor surveillance shall deter potential misadventure. In peacetime, it can promote regional stability by its persistent over-watch capabilities and during crises and wars it would enable rapid response across the spectrum of conflict by its ability to deter, contain, resolve or engage

---

13 Michael Krepon, "Weapons in the Heavens: A Radical and Reckless Option", Arms Control Today, November 2004, p. 10

14 J.R. Wilson, "The ultimate high ground", Armed Forces Joul71al, January 2004, p. 29.

and win.

- **Decision Making.** Space-based inputs augmented by conventional methods could be used for creating information databases on adversaries during peacetime, so that effort and resources are not disproportionately expended in acquiring basic inputs during hostilities. In addition, during crises and wars, they would enable effective decision-making by providing accurate and real (or at least near real-time) inputs vital to military decisions at strategic, operational and tactical levels.

- **Other Spin-Offs.** In addition to these applications, a number of other advantages accrue to the defence forces from space which are limited only by the ingenuity of the user and his 'employability-awareness' of the subject. For example, Met Sats can be used for enhancing battle-field employability and manoeuvre. Also the existing local user terminals for search and rescue can be applied for rescue of stranded combatants, ejecting aircrew, etc.

Imperatives of India's Defence Needs in Space

It is apparent that space-based systems provide vital capabilities to successfully execute national military strategy in addition to the overall grand strategy and have the potential to be used across a range of military operations at the strategic, operational and tactical levels of war. Secondly, information derived from air and space platforms would be vital for success in conflicts. Hence it would be imperative to attain a certain degree of 'information-dominance' in order to complement our conventional capabilities. Thus we need to enhance our conventional military prowess by harnessing available space capabilities and potential so as to comprehensively reciprocate to the spectrum of warfare being directed towards us and also limit (if not deny) our adversaries the opportunity to offset conventional military superiority by resorting to threats of WMD, or other forms of unconventional warfare. Thus there exists an emergent need for examining the options afforded by space in order to address the following aspects :

- Securing of our space assets and thereby ensuring uninterrupted national development.

- Coordination of military requirements and development of military space capabilities.

- Integration of space and conventional military capabilities.

Coordinating Agency for Space-based Military Operations

It is evident that today, space based assets offer a large number of war-winning capabilities like near instantaneous communications, continuous surveillance and highly accurate positioning. These capabilities provide a decisive advantage to the military. India has a formidable civilian space capability, while its military space capabilities are dismal. Over the years India has built up adequate capacity in space technology and our space assets are being exploited efficiently by the civil sector for a number of applications but the military use of space by India has been minimal, which has become focus of attention post-Kargil War. Present Indian efforts are grossly inadequate and uncoordinated in the absence of a central coordinating body. Indian military use of space is limited to procuring imagery from satellites like CartoSats and RISATs. It is yet to integrate its formidable civilian space capabilities into its war-fighting machinery or at least take steps to protect its assets in space. There has been no single agency earmarked for managing our military space programme resulting in a sub optimal and inadequate project management for exploiting our space capabilities. This lack of exploitation of our space capabilities for military use has resulted in significant gains being under-utilised in a military context. India ranks fourth in the world with regard to both national airpower and space capabilities. [15]

Thus the requirement of a centralised agency aimed at coordinating national military space effort in support of national military space objectives is imperative and inescapable. To this effect, a beginning has been made by India by creating a coordinating agency "Integrated Space Cell" (ISC) under HQ IDS which has taken up the task of formulating a Space Doctrine for armed forces

---

15 "An Aerospace Command", Vayu Magazine, March 2003, p. 16.

to cater for Offensive and Defensive Space Operations. However, in view of the expanding role space technologies play towards military applications and the larger role of the armed forces towards adapting these technologies towards war-fighting, the ISC needs to expanded to form a National Space Command (NSC), which can be another tri-services command like ANC and SFC, manned by specialist manpower from all three services. While no attempt is made herein to comprehensively list all the tasks which the ISC would fulfil, broadly the following requirements would be met:

- It would serve as a single point of contact for all military space applications and will support all space operations in due course.

- It would serve as a nodal agency for projection of space system/capability demand and its acquisition from DRDO.

- It could be entrusted with the responsibility of coordinating the requirements of all the three services, consolidating all existing and proposed space support applications, recommending compatibilities and feasibilities of such applications and scrutinising support requirements.

- It could be tasked to examine and recommend suitable defence oriented systems/sub systems, deployment of such systems and subject to approval, pursue implementation of such projects in co-ordination with DRDO and ISRO.

- Create a Space Corp comprising of personnel from all the three services and arrange for their training in consultation with DRDO, to provide them system specific hands on training.

- The ISC (or the upgraded NSC) shall take charge of the acquisition and adaptation of space technologies towards conduct of military operations in space, both offensive and defensive, so as to protect the national interest.

- Formulation of Space Doctrine for India along with a detailed organisational structure for the National Space Command and formulation of counter-space operational philosophies. The

Offensive and Defensive Counter-space Operations should be spelled out keeping in view the strategic requirements of India and protection of its space assets from any attack from adversaries, if any.

Counter-Space Operations

The counter-space operations are operations by which a nation's armed forces achieve and maintain space superiority, thereby providing freedom to attack as well as freedom from attack using space. Since today's armed forces are increasingly dependent on space in all types of warfare, it would be of interest for India to factor in Counter-space operations into its space doctrine. This will enable the armed forces to train and prepare for any war in the 21st century and beyond, wherein space based assets would be their mainstay. Since space infrastructure is largely unprotected, being the economic and strategic centre of gravities, they would surely be the most vulnerable assets, hence could be prime targets for attack in the event of war. As a result, counter-space operations are evolving rapidly given the reliance on space assets and indications that potential adversaries are beginning to exploit space both for their benefit and for military operations towards protecting national interests. The counter-space operations can be both offensive and defensive, which are dependent on robust space situation awareness (SSA). Counter-space operations are conducted across the tactical, operational, and strategic levels of war by the entire joint force (air, space, land, sea, information, or special operations forces). The counter-space operations can be distinctly divided into two categories, i.e., Offensive Counter-space Operation and Defensive Counter-space Operations. [16]

Space and air superiority will be a crucial first step in any military operation. Thus within the counter-space construct, any action taken to achieve space superiority is a counter-space operation. Few examples of counter-space operations include:

- Improving the commander's situational awareness and view of the battle space. Find, fix, track, target, engage, and assess space capabilities.

---

16   The US Air Force Doctrine Document 2-2.1 August 02, 2004pp 2-3

- Instituting appropriate protective and defensive measures to ensure friendly forces can continuously conduct space operations across the entire spectrum of conflicts

- Operations to deceive, disrupt, deny, degrade, or destroy adversary space capabilities.

Space Situation Awareness (SSA)

SSA is the result of sufficient knowledge about space-related conditions, constraints, capabilities, and activities—both current and planned—in, from, toward, or through space. Achieving SSA, supports all levels of planners, decision makers, and operators across the spectrum of terrestrial and as possible, the space capabilities operating within the terrestrial and space environments. SSA information enables defensive and offensive counter-space operations and forms the foundation for all space activities. It includes space surveillance, detailed reconnaissance of specific space assets, collection and processing of intelligence data on space systems, and monitoring the space environment. It also involves the use of traditional intelligence sources to provide insight into adversary space and counter-space operations.

Components of Space Situation Awareness. The components of SSA can be clubbed under the following activities:-

- Intelligence

- Surveillance

- Reconnaissance

- Command and Control

SSA Requirements and Tasks. Space situation awareness is more than surveillance of space. It is the command, control, communication and computers, intelligence, surveillance, and reconnaissance (C4ISR) and environmental data required for all space operations. SSA requires the following:-

- Use of intelligence sources to provide insight into adversary space doctrine, strategy, tactics, and operations.

- Surveillance of all space objects, activities, and terrestrial support systems.

- Detailed reconnaissance of specific space objects.

- Monitoring and analysis of the space environment.

- Monitoring and status of friendly, neutral, and adversary space assets, capabilities, and operations.

- Command, control and communications (C3) processing, analysis, dissemination, and archival capabilities used to accomplish these activities through a secured computerized network.

The tasks of SSA includes find, fix, track, and target, engage, and assess. Accomplishing these tasks ensures coherent battle-space awareness for planners, operators, and commanders.

Defensive Counter-space Operations

These operations preserve the nation's ability to exploit space to its advantage via active and passive actions to protect friendly space-related capabilities from enemy attack or interference. Friendly space-related capabilities include space systems such as satellites, terrestrial systems such as ground stations, and communication links. They are the key to enabling continued exploitation of space by protecting, preserving, recovering, and reconstituting friendly space-related capabilities before, during, and after an enemy attack. They may target an adversary's Counter-space capability to ensure access to space capabilities (e.g., an air strike against an active GPS jammer) and freedom of operations in space. [17]

Passive Measures. These measures limit the effectiveness of hostile action against space systems and are employed to launch counter attacks. Passive techniques for defensive counter-space operations include the use of camouflage, concealment, and deception,

---

17 The US Air Force Doctrine Document 2–2.1, Aug 02, 2004 pp 25-29

hardening of systems, and the use of dispersal. The objective of passive measures is to provide a layered defence and to withstand attack without warning.

- **Camouflage, Concealment, and Deception (CC&D).** It is most effective with terrestrial-based nodes. Certain types of ground-based components of space systems may operate under camouflage or be concealed within larger structures. These measures complicate identification and targeting.

- **System Hardening.** Hardening of space system links and nodes allow them to operate through attacks. Techniques such as filtering, shielding, and spread spectrum help to protect capabilities from radiation and electromagnetic pulse. Physical hardening of structures mitigates the impact of kinetic effects, but is generally more applicable to ground-based facilities than to space-based systems due to launch-weight considerations. Robust networks, hardened by equipment redundancy and the ability to reroute, ensures operation during and after information operations attack.

- **Dispersal of Space Systems.** For space nodes, dispersal could involve deploying satellites into various orbital altitudes and planes. For terrestrial nodes, dispersal could involve deploying mobile ground stations to new locations.

Active Measures. Active measures for Defensive Counter-space Operation may involve actions to avoid or remove hostile effects. Physical adjustments to the nodes and links of space systems, such as a maneuver or frequency change, may avoid hostile effects. Use of conventional or special operations forces may stop an adversary's counter-space attack. The key to these active measures is early detection and characterization of the threat in order to determine the most effective countermeasure.

- **Maneuverability/Mobility.** Satellites may be capable of maneuvering in orbit in order to deny the adversary the opportunity to track and target them. They may be repositioned to avoid directed energy attacks, electromagnetic jamming, or kinetic attacks from anti-satellite weapons (ASATs).

Today, maneuver capability is limited by on-board fuel constraints, orbital mechanics, and advanced warning of an impending attack. Furthermore, repositioning satellites generally degrades or interrupts their mission. The use of mobile terrestrial nodes complicates adversarial attempts to locate and target command and mission data processing centers. However, movement of these nodes may also impact the system's capability, as they must still retain line of sight with their associated space-based systems. Though the use of mobile technology is expanding, many of today's ground-based systems are not mobile, making physical security measures essential.

- **System Configuration Changes.** Space-based and terrestrial nodes may use different modes of operation to enhance survivability against attacks. Examples include changing RF amplitude and employing frequency-hopping techniques to complicate jamming and encrypting data to prevent exploitation by unauthorized users.

- **Suppression of Adversary Counter-space Capabilities (SACC).** It neutralizes or negates an adversary offensive Counter-space system through deception, denial, disruption, degradation, and/or destruction. SACC operations can target air, land, sea, space, special operations, or information operations in response to an attack or threat of attack. Examples of SACC operations include (but are not limited to) attacks against adversary anti-satellite weapons (before, during, or after employment), intercept of anti-satellite systems, and destruction of RF jammers or laser blinders.

- **Redundancy.** Redundancy may be incorporated into space-based or terrestrial capabilities or within a link itself. Redundancy in equipment components allows continued operations of specific platforms in the event of onboard hardware or software malfunction. Full systems may have redundancy through the use of on-orbit satellite spares, or use of alternate commanding, tracking, and relay stations. Link redundancy can be achieved through the use of alternate frequencies for command or mission information along with data multiplexing techniques.

- **Reconstitution.** Reconstitution involves actions to restore operations after an attack. Reconstitution may involve repairing equipment that has been degraded or it may entail deploying new space and terrestrial platforms to replace combat losses. Reconstitution of satellite constellations requires responsive space-lift (tactical launch capability), availability of replacement spacecraft, and properly trained personnel to launch and operate the systems at a short notice.

Defensive Counter-space Resources and Forces

The following are some of the forces and weapon systems that could be used, if and when available, to support Defensive Counter Space operations:

- **Single Integrated Space Picture.** This would provide an accessible picture of global and theater space capabilities, threats and operations to commanders, planners, and combat forces, covering the full spectrum of friendly, adversary, and third party space systems. In addition, it would provide comprehensive peacetime and wartime situation awareness, fusing information collected on all space systems, their ground, air, and space links and nodes to include their capabilities, status, vulnerability, and users.

- **Physical security systems.** This will provide security and force protection for critical ground facilities and equipment. A complementary mix of technology and security forces can effectively and efficiently mitigate specific threats in an ever-changing environment. When properly deployed and utilized, physical security systems can represent an effective deterrent and provide aggressive defence against terrestrial node, attack and sabotage.

- **Air defence assets.** They are capable of protecting launch and terrestrial nodes from air or missile attack. If threatened, commanders should consider deploying air defence assets such as fighter aircraft, surface-to-air missiles, and/or antiaircraft artillery to protect critical space assets (e.g., facilities and infrastructure). A sound air defence may deter an adversary and most certainly will be instrumental in defending our forces and assets if an attack is attempted.

- **Attack detection and characterization systems.** It can detect space system attacks and provide information on the characteristics of the attack, especially if the source and/or capability of the attack is unknown or unexpected. These systems will support locating the source of the attack and the type of weapon used in the attack. They may be ground-, air- or space-based and either integrated with systems they protect or used in a stand-alone capacity. Having our adversaries aware of these capabilities may influence their decision and act as an effective deterrent.

- **Survivability.** This feature of a satellite will ensure critical space systems continue to operate both during and after attack. Examples include spacecraft system hardening, redundant systems (both on spacecraft and in ground stations), spacecraft maneuverability, ground station mobility, and jam-resistant communication links. Known survivability measures may deter an adversary from attacking our space capabilities.

- **Operations security (OPSEC) and information assurance (IA).** Both activities can protect our space systems by limiting the availability of information on their operations, capabilities, and limitations to our adversaries. IA protects critical computer systems from intrusion and exploitation. Guiding adversaries' actions can successfully deter effects on our space services, but OPSEC and IA operations are primarily focused on defending our assets from attack.

- **Conventional and Special Forces Operations,** They are conduct of defensive counter-space operations through their ability to attack adversary counter-space capabilities. A demonstrated capability and willingness to counter their counter-space capabilities may deter an adversary from attacking US/friendly space capabilities.

Offensive Counter-Space Operations

These operations preclude an adversary from exploiting space to their advantage. They may target an adversary's space capability (space systems, terrestrial systems, links, or third party space capability), using a variety of permanent and/or reversible means. The "Five Ds" —deception, disruption, denial, degradation, and destruction—describe the range of desired effects when targeting an adversary's space systems. As adversaries become more dependent on space capabilities, Counter-space operations have the ability to produce effects that directly impact their ability and will to wage war at the strategic, operational and tactical levels. Denying adversary space capabilities may hinder their ability to effectively organize, coordinate, and orchestrate a military campaign. For example, if adversaries reconstitute their command and control (C2) capabilities via satellite communications (SATCOM) after their ground-based communications network has been destroyed by precision bombing, offensive counter-space operations may be employed in conjunction to reduce or eliminate their ability to communicate with their forces.[18]

The "Five D's"—deception, disruption, denial, degradation, and destruction—are the possible desired effects when targeting an adversary's space capability.

- **Deception** employs manipulation, distortion, or falsification of information to induce adversaries to react in a manner contrary to their interests.

- **Disruption** is the temporary impairment of some or all of a space system's capability to produce effects, usually without physical damage.

- **Denial** is the temporary elimination of some or all of a space system's capability to produce effects, usually without physical damage.

- **Degradation** is the permanent impairment of some or all of a space system's capability to produce results, usually with physical damage.

---

18 The US Air Force Doctrine Document 2–2.1, Aug 02, 2004 pp 31-34

- **Destruction** is the permanent elimination of all of a space system's capabilities to produce effects, usually with physical damage.

Offensive Counter-space Operations Targets

Offensive counter-space operations seek and attack targets in three general categories: space nodes, terrestrial nodes, and links. Space nodes may include satellites, space stations, or other spacecraft. Terrestrial nodes include land, sea, or airborne equipment and resources used to deploy, enable, interact with, or otherwise affect the space node. Communication links tie nodes together, and pass information between them. Understanding space capability as a system of nodes and links, enables one to determine the best ways and means for affecting adversarial capability.[19] The following are examples of offensive counter-space targets:

- **On-orbit Satellites.** Satellites are on-orbit assets consisting of a mission sensor and a satellite bus. The mission sensor provides raw data, which is usually sent to a ground station for processing. The satellite bus carries the mission sensor and provides it power, thermal control, and communications. Offensive counter-space operations may target the mission sensor or the satellite bus. For example, a laser may deny, disrupt, degrade, or destroy certain types of sensors. Kinetic anti-satellite weapons, on the other hand, usually target the satellite bus for physical destruction.

- **Communication Links.** Space systems are dependent on RF and/or laser links to provide communication between space and terrestrial nodes (satellite to ground station or satellite to user), between terrestrial nodes (ground station to users), and between satellites (satellite to satellite). On-orbit satellites and ground-based satellite control stations/users send data up and down the link. In the up-link, command and control data tasks satellite mission payloads and subsystems. In the downlink, mission payload and satellite state-of-health data are sent to a ground station for processing. The ground station, after processing, often sends the mission data to the

---

19 Ibid pp 31-34

users via SATCOM for exploitation. In the case of SATCOM systems, data may be directly up-linked and then down-linked between users. Most space systems are ineffective without communication links.

- **Ground Stations.** Ground-based systems perform satellite command and control and mission data processing. Ground stations are normally permanent structures that may represent a single point of failure in a space system. Mobile ground stations can also be used to command and control a satellite, but may have no ability, or a limited capacity, for processing satellite mission data.

- **Launch Facilities.** The ability to place satellites into orbit is the first step to space access; fundamental to the ability to operate and maintain space-based capability. Whether this capability is indigenous, or provided by a third party, it is the only means to deploy satellites to space and represents a primary choke-point for interdicting an adversary's efforts to augment or reconstitute space forces.

- **Command, Control, Communication, Computer, Intelligence, Surveillance, and Reconnaissance (C4ISR) Systems.** C4ISR systems are critical to the effective employment of forces and assets. Destruction of such systems would substantially reduce the enemy's capability to detect, react, and bring forces to bear against friendly forces. Attacking C4ISR systems may contribute to offensive counter-space operations but may also contribute to strategic attack or counter-air operations, depending on the intended effects.

- **Third Party Providers.** An adversary may gain significant space capabilities by using third party space systems. Using diplomatic or economic means to deny an adversary access to these third party (commercial or foreign) space capabilities will generally require the assistance of other governmental agencies.

Offensive Counter-space Resources and Forces

The effectiveness of offensive counter-space operations to affect the array of targets previously listed depends on the availability and capabilities of certain resources and systems. The choice of system depends upon the situation, threats, weather, and available intelligence. Whenever possible, use systems and methods which minimize risk to friendly forces. For example, an aircraft employing standoff weapons may provide the same effect as a Special Forces team on a direct mission, with less risk to friendly forces. The following are some of the forces and weapon systems that could be used, if and when available, to conduct offensive counter-space:

- **Aircraft.** Friendly aircraft provide non-kinetic and kinetic capabilities against surface targets associated with an adversary's space capabilities. For example, electronic attack platforms (manned and remotely piloted aircraft) could affect the links of an adversary's space system employing stand-off and stand-in techniques. By attacking terrestrial nodes, aircraft may disrupt, deny, degrade or destroy an adversary's ability to control their satellites or deliver space effects.

- **Missiles.** Missiles may be employed against a variety of an adversary's space capabilities including launch facilities, ground stations, and space nodes.

- **Special Operations Forces (SOF).** They can conduct direct attacks against terrestrial nodes or provide terminal guidance for attacks against those nodes. Additionally, SOF may be used to provide localized jamming of an adversary's links.

- **Offensive Counter-space Systems.** These systems are designed specifically for OCS operations, such as a counter satellite communications capability, designed to disrupt satellite-based communications used by an adversary or a counter surveillance reconnaissance capability, designed to impair an adversary's ability to obtain targeting, battle damage assessment, and information by denying their use of satellite imagery with reversible, non-damaging effects.

- **Anti-satellite Weapons (ASATs).** ASATs include

direct ascent and co-orbital systems that employ various mechanisms to affect or destroy an on-orbit spacecraft.

- **Directed Energy Weapons (DEWs).** DEWs, such as lasers, may be land, sea, air, or space based. Depending on the power level used, DEWs are capable of a wide range of effects against on-orbit spacecraft, including heating, blinding optics, degradation, and destruction. Under certain circumstances, lasers may also be effective against space launch vehicles while in-flight.

- **Network Warfare Operations.** Many offensive counter-space targets, particularly elements of the terrestrial node, may be affected by various IO techniques such as malicious codes, electronic warfare, or EMP generators. Some IO techniques afford access to targets that may be inaccessible by other means.

- **Electronic Warfare Weapons.** Radio Frequency jammers may be used to disrupt links.

- **C4ISR Systems.** These systems include early warning and surveillance systems, satellites, radar, identification systems, communications systems, and surface/air/space based sensors. These systems enhance offensive counter-space operations by providing early warning, intelligence, targeting, and assessment data, as well as C2 of friendly forces.

- **Surface Forces.** The ability to occupy and secure key areas, as well as the lethality of supporting surface forces can achieve significant Counter-space effects. For example, surface forces can attack a satellite control station in order to disrupt, degrade, or destroy an adversary's space capabilities.

CHAPTER - V

CHINA'S RESURGENCE IN SPACE: MUSCLE TO ITS MILITARY POWER

The People's Republic of China (PRC) has made significant advances in its space program and has now emerged as a formidable space power. The credit for this goes to the senior leaders of PRC, who established space as a 'national priority' and allocated significant resources toward enhancing its space-related technology base. Apart from adding to China's comprehensive national power, the achievement of PRC in the field of space, especially its manned program enhances its national prestige and draws international attention to the country's expanding technology base. China's investments in space also serve as a stimulant to its economic growth in terms of space commerce.

China's space ambitions are basically peaceful in nature and are meant for the societal good. However, since space technologies can also be used to support military operations, the PLA is rapidly improving its space and counter-space capabilities in order to advance the interests of Chinese Communist Party (CCP) and also to counter the perceived challenges to its sovereignty and territorial integrity. Space capabilities enable the PLA to conduct military operations at increasingly greater distances from Chinese shores challenging US dominance in the Asia-Pacific region. The PRC's dedicated efforts over the last two decades have borne fruit in the fields of advanced space technologies, precision strike assets integrated with dedicated space-based surveillance, integrated air and space domain, and reliable communication architecture. This will help China secure its national interests and will also enable it to negate the US dominance in the world.

China has consolidated its space technologies and has been able to make considerable progress in advancing its military space capabilities. PLA's robust and growing space-based sensor architecture and its ability to transmit reconnaissance data to ground stations in near real time, enables PLA to conduct long-range precision strikes with growing lethality and accuracy. China is today focusing on increasingly high resolution, dual-use space-based electro-optical, synthetic aperture radar, and electronic intelligence satellites for surveillance and targeting. Based on the improved space architecture and network of sensors in space and development and production of dedicated military communications satellites and space based data link satellites, the PLA will be able to transmit high volumes of data from from/through space to a numerous users that support its global operations. The independent Beidou Navigational System will further improve accuracy and precision of Chinese military operations by giving them an alternative to the US GPS navigation system.

In China, it is the PLA General Armaments Department (GAD) which is responsible for the space systems acquisition, including technical design, research and development (R&D), manufacturing, and space launch services. GAD is improving its ability to integrate the technical expertise that is available throughout China's defence industrial infrastructure and academic community. Today, China's increasingly sophisticated R&D and a dedicated industrial establishment is capable of supplying the PLA with the required military space systems. China Aerospace Science and Technology Corporation (CASC) and China Aerospace Science and Industry Corporation (CASIC) are the two most significant space organizations of China, which have been able to put the country on a path of development and security, both accrued as a result of development and exploitation of space technologies. In addition, the China National Space Administration (CNSA) facilitates international exchanges and cooperative programs with other space-faring nations and is supposed to be an equivalent of NASA of the US.

The PRC has thus prioritized international space-related interactions in order to further its political, scientific, technological, and economic goals. In this regard China enjoys a very close cooperative relationship with space authorities and engineers from Russia, other CIS and European countries. Also, the PLA has been investing in a wide range of R&D activities which include satellite communications monitoring systems, electronic countermeasure systems to disrupt an opponent's use of space-based systems, and developing the capability for physical destruction of satellites in orbit. China has eventually been successful in manned space platforms, reliable space launch vehicles, and satellites. Concurrently it has made substantial progress in commercial utility of its space technology by taking up cost effective international commercial launch services. PRC's space program today supports economic development through subsidized modernization of China's high technology industries, contributing to natural disaster warning and response, and developing commercial applications of space technology. Though Chinese white papers in 2000, 2006 and 2011 highlights the CCP's vision for enhancement of its space technologies towards securing its national interests, which stresses the peaceful use of space. Yet the ASAT test by China in 2007 leaves little doubt about its militarization and weaponisation programmes and dual use capabilities of the available space technologies, which increases the capacity of the PLA to project military power vertically into space and horizontally beyond its immediate periphery. The freedom of action in space offers the PLA potential military advantages on land, at sea, and in the air. Though at present the Chinese space and counter-space capabilities are yet to be integrated, synergized and operationalised, yet the PLA is rapidly improving and consolidating its space and counter-space capabilities in order to support CCP interests and defend against perceived challenges to the sovereignty and territorial integrity. The immediate gain to the PLA will be that the space assets will enable extended range precision strike operations intended to deny the US access to or an ability to operate within a contentious area of the Indo-Pacific region.[1]

---

1 Wayne A. Ulman, "China's Emergent Military Aerospace and Commercial Aviation Capabilities," Testimony before the U.S.- China Economic and Security Review Commission, 20 May 2010, http://www.uscc.gov/hearings/2010hearings/written_ testimonies/10_05_20_wrt/10_05_20_ulman_statement.php.

Space: China's National Priority

China is the first developing nation which has successfully developed "full-spectrum" of aerospace technologies. Compared to the other nascent space players like India, Japan and Brazil, who started their space journey almost during the same era, but the mercurial advancement gained by China in space technologies, reflects the dedicated commitment of Chinese leadership. The commercial gains which are linked to the dual use space technology has ensured that the economic gains accrued out of the commercial spin offs of the successful space technologies are used to support the space programme and its military utilities.

The pace of space development in China can be gauged by the fact that since the time Mao Tse Tung initiated China's space program in October 1956; within four years China launched its first rocket on 05 Nov 1960, becoming the fourth country after Germany, the US and the erstwhile Soviet Union, to enter space. The transformation has been such that today China routinely launches space satellites for even the European companies and the US corporations, thereby increasing its share in the global space launch market. The credit for the rapid development of the space technology in China therefore goes to the Chinese leadership, which placed the aerospace development as one of the top priority of the nation.[2] The results are there for others to see, as to how China has today almost overshadowed the American Space program by overtaking the US in launching higher number of satellites in 2012.

An Overview of Chinese Strategy and Approaches to Space Exploitation

China began investing strategically in space science and technology from 1956 onwards with Soviet assistance. Over the next four decades, it embarked on a number of landmark space programs and succeeded in launching its first satellite in 1970s and developing its retrieving technology in the 1980s; it also conducted manned space

---

2 **Dr. Andrew S. Erickson** is an associate professor in the Strategic Research Department at the U.S. Naval War College and a founding member of the department's China Maritime Studies Institute (CMSI). He is also a Fellow in the National Committee on U.S.-China Relations' Public Intellectuals Program.

exploration in the 1990s. In the 2000s, PRC succeeded in orbiting the moon. It has launched a space module Tiangong-I, which recently hosted its first team of Chinese taikonauts, who successfully conducted 'manual docking' of their space ship Shenzhou -9. It is now testing its heavy-class launch capability, which will eventually lift heavy components of its space station which is likely to be operationalised by 2020.[3] They are also preparing for a manned landing mission on the moon in this decade.[4] The underlying intent for both, the Chinese leadership and its strategists is to ensure space prowess to PLA which will help the nation to attain the desired position in the New World Order. China not only has become a leading space power amongst the developing nations, but has also reached to some parity with the US and Russia in many of the niche space technologies.

China's space policy guidelines and projections, state regulations on Science and Technology units supporting space projects, and three successive government white papers have clearly spelt out a space strategy of setting up its space missions. These missions include manned space projects; lunar probing; high-resolution earth observation, which are relevant to China's island disputes with other nations; and assessing the feasibility of lunar and deep space exploration, including a mission to Mars towards exploring and mining of gold, platinum, and rare earth metals. Towards this, PRC has set up various agencies to administer the different aspects of its space endeavors—civil, military, and commercial. A nationwide and competent infrastructure has been build up in terms of research and design, manufacturing, launch, and service to support the space programmes.

The Chinese government has also made it clear that the purpose of its space development is threefold: to explore outer space so as to expand the understanding of Earth and the universe; to peacefully tap into outer space to promote human civilization and social progress and to benefit all of humankind; and to meet the needs of China's

---

3 People's Daily on Line, 26 Jun 2013, "China plans to launch Tiangong-2 space lab around 2015" accessed through http://english.peopledaily.com.cn/202936/8300039.html

4 ShenDingli in "China's Perspective on Space Security"

economic construction, scientific and technological advancement, national security and social progress, while enhancing national science and cultural quality, defending state interests and lifting overall national strength.[5]   With these principles in mind, China has put forward its guidelines for space strategy, which involves innovativeness, selectivity, support, and leadership. The Chinese leadership emphasises that space strategy has to be leveraged towards overall national development. Thus, in order to strengthen the nation, space technologies have been strategically viewed as necessary to develop China's capacity in the economic, science and technology, defence, and social cohesiveness. Keeping in view the benefits the space programme is accruing to the nation, it is imperative for the Chinese leadership to be committed to a long-term space development programme. The two distinct aspects of the Chinese space programme, commercial and strategic, are discussed below.

Commercial Perspective of the Chinese Space Programme

China has started its space programme for the benefit of its people. Public welfare and utilities were the key features important to their national interests. Commercial utility was instrumental in accelerating the building up its space capabilities and a robust space industry to support its space programme. With its strength of central planning and, organized command structure, China relentlessly focused on the basic research and aspired to lead research in selected fields, as well as on implementing major projects to enable a leap in space engineering. It made significant gains in satellite manufacturing, succeeded in retrieving satellites and also in multiple-satellite launching, cryogenic engines for high lift rocket launchers and cryogenic fuel, and add-on boosting technologies for heavy lift rockets. The Chinese have boldly demonstrated their space capabilities time and again by producing six series of satellites: retrievable remote sensing, communication, meteorology, science and technology, earth resources surveying, and navigation and positioning. It has also got a project of micro satellite constellations for environmental and disaster forecasting which has obviated the

---

5 Ashley J Tellis and Sean Mirski in "Crux of Asia- China, India and the Emerging Global Order" pp 166-167

need for heavy cumbersome satellites. It has also made significant progress in micro-satellites. Recently, Chinese authorities made public that the country has invested $6.2 billion on its manned space program and this programme has already given a return of around $16 billion.[6] This shows the immense commercial utility of the Chinese space programme, which is expected to increase the economic yield three times by 2015.

China has been able to commercially exploit its new generation of rocket carriers for the business of commercial launch of satellites. They have also been able develop a liquid-oxygen/kerosene engine with a thrust of up to 120 tons, as well as an oxy-hydrogen engine with a thrust of 50 tons, in preparation for its next stage of heavy lifting. In order to boost its heavy lift capabilities to support construction of a permanent space station in space, it is constructing its fourth space launch center in Hainan. It has already demonstrated its capabilities towards manned space missions, and it is pushing for deep-space exploration starting with moon orbiting and landing, which is likely to be followed to Mars. It will further continue conducting major projects on its manned space program, lunar exploration, high-resolution earth survey system, and new rocket carriers. In particular, it has achieved spacecraft docking, both automated and manual and is now preparing for launching its R&D space science research program on a space laboratory with periodic manned operations. In order to transport supplies to its futuristic space stations to sustain manned projects, China plans to lift up to 25 tons of payload to near-Earth orbit and 14 tons of payload to earth synchronous transfer orbit from its new launch centre. It is also conducting research on space telescopes, space astronomy, astrophysics, and space life science, among other projects; strengthening surveillance of space debris; and creating an early warning system for space monitoring.

The latest money spinner for China is the Beidou Navigation System, which has been offered for commercial use, as of now in the Asia Pacific region. Today a total of four countries apart from China have integrated Beidou system commercially which include Laos,

---

6 "China's Present Cost of Total Manned Space Program at RMB $39b," www.china.com.
cn/news/2012-06/24/content_25723767.htm, June 24, 2012.

Brunei, Thailand and new entrant Pakistan.[7] However, in 2020, when this system will be able to provide world-wide coverage, the commercial outcome is likely to rise exponentially. According to a web based space portal GPS Daily, "China's domestically-produced navigation system aims to take 70 - 80% of the now GPS-dominated domestic market by 2020." [8]

Strategic Perspective of the Chinese Space Programme

An outstanding feature regarding the Chinese space programme has been that along with its civilian space programme, it has utilized its space technology towards securing its national interests. China has thus ensured that its space assets compliments the nuclear capabilities in order to secure its national interest. While nuclear capabilities, to some extent, manage the balance of power on land, China's space capabilities would extend the balance to outer space, serving the purpose of stabilizing international relations among nations. Despite the fact that China today is the leading space power in Asia and perhaps competing with both the US and Russia, it has never projected space hegemony through its space technologies. Instead it has so far signed all the UN-sponsored international agreements for the peaceful use of space. It has strongly supported in the UN a treaty for the Prevention of an Arms Race in Outer Space. However, after witnessing the militarization of space by the US and its utilities to change the tide of war in Iraq, Afghanistan, Kosovo and Libyan conflicts, China realized that it would never be able to match the American technological superiority and strength. Thus, China decided to weaponise space by testing an ASAT weapon in 2007, so as to gain asymmetric advantage over the much superior American troops and technological superiority.

Apart from developing the ASAT capabilities, the US also suspects China to have painted its spy satellite by firing a ground-based laser beam. An analysis of the satellites launched by China, it has been confirmed that it has deployed several reconnaissance and

---

7 News report, "Pakistan... Shuns GPS, Goes For Beidou" reported by SATNEWS, a web based portal accessed through http://www.satnews.com/story.php?number=60069131

8 News published by web base space portal GPS Daily titled "China eyes 70-80 % market share for its GPS rival" on 31Dec 2012 accessed through http://www.gpsdaily.com/reports/China_eyes_greater_market_share_for_its_GPS_rival_999.html

communication satellites in space to fulfill its C4ISR requirements to counter any threats to its national security. In the same league, exactly three years after its ASAT test, China carried out a successful mid-course interception test of a ballistic missile on January 11, 2010, thereby giving rise to speculations about China's military space strategy. Speaking on the occasion of the 60th anniversary of the Chinese Air Force in 2009, General Xu Qiliang, the then Air Force commander pointed out that,

> *"As the major powers are shifting toward air-space integration under the circumstances of informatization, China has also established its own air force strategy as "air-space, and offence/defence integration."[9]*

Thus it can be seen that China does not accept the US as a sole leader to capitalize and manipulate space for its economic and strategic gains. It still feels that space is a "global common" which needs to be shared by every nation which has space capabilities. Thus the US dominance of space is unacceptable to China. Hence, China is trying to develop the asymmetric technologies to counter the American threats from space technologies, so as to counter balance the power whenever possible in due course. As the Chinese economy is growing leaps and bounds, it will surely surpass the US economy and the economic gains will thus be used to consolidate the military might in space to challenge and counter the 'space superiority' enjoyed by the US. The result of the Chinese pursuits in space is visible in terms of the fact they have devised ways and means to counter President Obama's rebalancing strategy or the "Pacific Pivot." This has resulted into renewed and intense activities in South China Sea and East China Sea, both diplomatic and military. This unhealthy rivalry and competition could lead to an armed conflict, if not checked and resolved in time.

Foreign Co-operation and Collaborations in the Chinese Space Program

The PRC has prioritized international co-operation in space

---

9 "China's Air Commander: Chinese Air Force Shall Strengthen Air-Space War-Fighting Capability," Nov 6, 2009 accessed through http://military.people.com.cn/GB/1076/52966/10327787.html.

programme right from the beginning. The intent was to consolidate and advance its national goals. In order to focus on space-related high end technologies and other applications of space systems, CNSA has formed multilateral and bilateral partnerships with a wide range of international partners. The bilateral exchanges include cooperative relationships between universities and research centers with counterparts in the United States, Taiwan, Europe, Russia, and the Ukraine. For instance, a Chinese technical journal credited a major U.S. university with helping them overcome a specific technical problem related to space interceptor KKV development.[10] Collaboration with Ukraine's Academy of Sciences helped Chinese civilian researchers develop advanced ablative heat resistant materials for maneuvering boost-glide re-entry vehicles.[11]

However after progressing to respectable standards in space technologies, they have now started collaborating with countries which are trying to make forays into space and harness the existing space technologies towards their economic progress. In recent years, China has launched satellites for Venezuela, Pakistan and Sri Lanka. The recent launch by China for foreign partners was micro satellites for Turkey, Argentina and Ecuador through its Long March-2D carrier rocket on 26 Apr 23.[12] According to a report, China's Great Wall Industry Corp has till now launched 35 rockets carrying a total of 41 satellites for foreign clients.[13]

China's Command & Control Structure for Space Exploration & Exploitation

The PLA's General Armament Department which manages the Chinese space programme is also responsible to the Central Military

10 Chen Dingchang, Wan Ziming, Lin Jin, and Liu Decheng, "Review of U.S. SPIE'98 Aerodynamics and Opto-Electronics Conference" May 1999, at http://www.space.cetin.net.cn/docs/mp9905/mp990513.htm.

11 The Academy of Sciences website at http://www.ipp.kiev.ua/about/conne_e.html.

12 News report by SATNEWS On 29 Apr 2013 titled "China...A Triple Micro Play (Launch)" accessed through http://www.satnews.com/story.php?number=359404405

13 News in The Economic Times titled "China successfully launches Turkish earth observation satellite" on 19 Dec 12 accessed through http://articles.economictimes.indiatimes.com/2012-12-19/news/35912698_1_long-march-2d-carrier-carrier-rocket-china-aerospace-science

Commission (CMC) for establishing defence and space acquisition policies, developing technical solutions to fructify operational requirements, and overseeing defence industrial research, development, and manufacturing. Thus the successful execution of national and military space acquisition policies is the core activity of GAD, which is well established agency today. It oversees a national level science and technology advisory committee, a number of administrative departments, and operational space units. In order to have effective control over the development and operationalisation of space technologies, the GAD is sub-divided into the following sub groups:

The GAD Science &Technology Committee

The GAD S&T Committee consists of expert working groups that advise CMC members and civilian authorities on long term technology acquisition planning and space policy and operations. At least 20 national level technology working groups, supported by defence R&D laboratories around the country, leverage and pool resources to review progress, advise national leaders on resource allocation and focus and prioritise resources to overcome technological bottlenecks.

GAD Administrative Departments

GAD also consists of as many as 10 second level departments responsible for various facets of force modernization, space planning and programming, and space operations. A GAD Space Equipment R&D Center appears to serve as an interface with space system users. The Comprehensive Planning Department appears responsible for overall space-related modernization planning and policy. Space architecture development is managed by the GAD Electronic and Information Infrastructure Department, which is China's leading authority for planning, programming, and budgeting for PLA "informatisation" development. It consists of at least four bureaus and one program office. The Aerospace Equipment Bureau is responsible for charting the PLA's future space-based communications and surveillance architecture and may manage R&D and manufacturing contracts with the space and missile industry. The Department also has program management functions, such as the Beidou Program Office, also known as the China Satellite Navigation System Management Office.

China's Space Command

The GAD Headquarters also functions as an operational command responsible for space launch, tracking, and control. The space programme is thus managed by the GAD Chief of Staff, who also operates as the head for China Satellite Launch and Tracking Systems Department (CLTC), a department which oversees China's space launch operations and management of all launch centers. The pace of the Chinese space programme can be appreciated from the fact that between 1970 and early 2012, CLTC launched a total of 157 satellites for domestic as well as international customers.[14]

China's Satellite Launch Centres

China has launched its entire range of satellites from its own launch centres which are based at following locations:

- **Jiuquan.** This Satellite Launch Center supports LM-2C, LM-2D, and LM-4 launch of satellites into low earth orbit, as well manned space missions on the LM-2F. It also handles the ballistic and land attack cruise missile testing.

- **Taiyuan.** This Satellite Launch Center functions as China's primary platform for satellite launches into sun synchronous orbit. It is also a key facility for the testing of medium and intermediate range ballistic missiles.

- **Xichang.** This Satellite Launch Center is China's primary platform for launch of satellites into geosynchronous orbit (GEO). It has a capacity to launch around08 to 10 satellites a year. This centre came into limelight for the Chinese ASAT test 11 January 2007 as a kinetic kill vehicle (KKV) against an aging Chinese weather satellite.[15]

---

14 *Chinese Launchers and COMSATs: Development & Commercial Activities*, briefing by Fu Zhiheng (Vice President, China Great Wall Industry Corporation) for World Space Risk Forum, Dubai, February 28-March 1, 2012.

15 *Ibid*

- **Wenchang.** This Space Launch Center on Hainan Island is under construction and will eventually serve as a base for launches with heavy payloads associated with China's manned space program. The launch vehicle will be transported to Hainan via ship from new manufacturing facilities in Tianjin, rather than rail.[16] This center will be the most modern of the launch centers which will be the hub of Chinese space activities for establishing its Space Station by 2020.

China's Space Tracking Control and Surveillance Network

The GAD has developed an elaborate and well-established infrastructure for space tracking, control, and surveillance. This tracking control and surveillance center has it's headquarter at Weinanin in Shanxi Province and functions as a space and missile surveillance center. It plays a very important role in monitoring and identifying debris and other objects in space. The Chinese space tracking network consists of a center in Xian, fixed land based sites, at least one mobile system, and four Yuanwang tracking ships capable of operating throughout the Pacific, Atlantic, and Indian Oceans. Apart from these centers which enable space tracking and control for the Chinese satellites and space objects including debris monitoring, GAD also operates a large number of foreign tracking and control locations. Thus all the Chinese space operations are managed from the Beijing Space Command and Control Center under GAD.

Chinese Launch Vehicles

It may be noted that the Long March (LM) launch vehicle has its roots in the country's ballistic missile program, specifically the Dongfeng-4 (DF-4) and Dongfeng-5 (DF-5) intercontinental ballistic missile (ICBM) systems. As of date there are four basic series of LM liquid-fueled launch vehicles which delivers payloads to orbits at varying altitudes and inclinations around the earth. The LM-1became the first Chinese launch vehicle to successfully launch a satellite into low earth orbit in April 1970. The LM-2 series has

---

16 XinDingding, "New Carrier Rocket Series to be Built," *China Daily*, October 31, 2007, at http://www.chinadaily.com.cn/china/2007-10/31/content_6217880.htm.

been used for delivering both remote sensing and communications satellites from Jiuquan and Xichang Space Launch Centers. The LM-2F is China's most powerful launch vehicle to date, which has the capability to launch more than 8,000 Kg into low earth orbit. Sharing the same first and second stage as the LM-2C, the LM-3 series integrates a cryogenic third stage that has been used for boosting heavier payloads into space from Xichang Space Launch Center.

Other launch vehicles include the LM-2D and LM-4 series, which have transported remote sensing, weather, and other payloads in sun-synchronous orbit from Taiyuan Space Launch Center. The LM-2D has launched payloads into both low earth orbit and sun synchronous orbit from Jiuquan Space Launch Center.[17]

Since 2008, China has been investing resources into a new generation of launch vehicles, including the LM-5, LM-6, and LM-7. The LM-5 is said to be designed to lift a 25 tons payload to low earth orbit (LEO), or a 14 tons payload into geostationary transfer orbit (GTO). Further, the R&D for LM-6, which is expected to be a smaller launch vehicle capable of boosting 500 kg into orbit, began in Sep 2009. The LM-7 is designed to place a 5.5 ton payload into a sun-synchronous orbit at an altitude of 700 km.

China's Space Industrial Infrastructure

China has got two state owned enterprises to support its space and missile industry, China Aerospace Science and Technology Corporation (CASC) and China Aerospace Science and Industry Corporation (CASIC). Both of them are organized on the lines of a corporate entity and have got under them various business divisions which are referred as 'academies'. Each academy focuses on a core competency, such as medium range ballistic missiles, short range ballistic missiles, ICBMs and satellite launch vehicles, cruise missiles, and satellites. CASC/CASIC academies are organized into research institutes focusing on specific sub-systems, sub-assemblies, components, or materials; and their testing and manufacturing facilities.

---

17 "The LM-2D," China Great Wall Industry Corporation website, April 1, 2010, at http://www.cgwic.com/LaunchServices/LaunchVehicle/LM2D.html.

China Aerospace Science and Technology Corporation (CASC)

CASC is responsible for development and manufacture of space launch vehicles, strategic ballistic missiles, satellites, and other space flight vehicles. Its functional business divisions specialize in ballistic missiles and space launch vehicles, large solid rocket motors, liquid fuelled engines, satellites, and related sub-assemblies and components. In the second phase of its development, CASC further consolidated its area of responsibility in inertial measurement units, telemetry, and missile-related micro-electronics, such as the high performance digital signal processors and field programmable gate arrays that are needed for long-range precision strike at high speeds and extreme temperature conditions. It is the CASC Science &Technology Committee which advises the State Council, Central Military Commission (CMC) and CASC leadership on space technology issues. The organization chart of CASC is highlighted below:

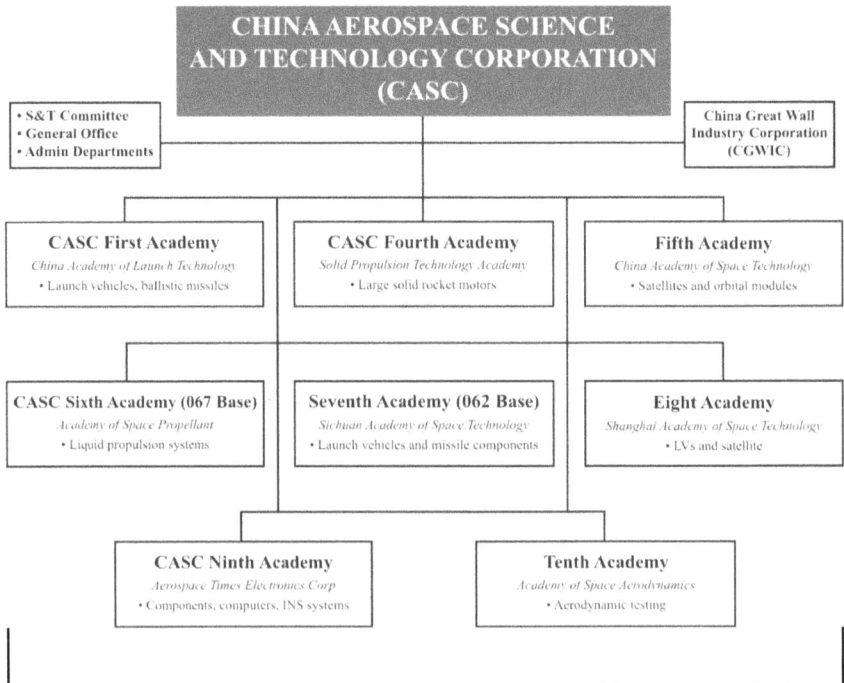

CHINA AEROSPACE SCIENCE AND TECHNOLOGY CORPORATION (CASC)

- S&T Committee
- General Office
- Admin Departments

China Great Wall Industry Corporation (CGWIC)

CASC First Academy
China Academy of Launch Technology
- Launch vehicles, ballistic missiles

CASC Fourth Academy
Solid Propulsion Technology Academy
- Large solid rocket motors

Fifth Academy
China Academy of Space Technology
- Satellites and orbital modules

CASC Sixth Academy (067 Base)
Academy of Space Propellant
- Liquid propulsion systems

Seventh Academy (062 Base)
Sichuan Academy of Space Technology
- Launch vehicles and missile components

Eight Academy
Shanghai Academy of Space Technology
- LVs and satellite

CASC Ninth Academy
Aerospace Times Electronics Corp
- Components, computers, INS systems

Tenth Academy
Academy of Space Aerodynamics
- Aerodynamic testing

Organization Chart of CASC[18]

---

18 "China's Evolving Space Capabilities : Implications for US Interests" a report by Mark A Stokes and Dean Cheng

The CASC functions through different academies which specialize in different aspect of space system. The details are as follows:

- **China Academy of Launch Technology (CASC First Academy).** The CASC First Academy, also known as the China Academy of Launch Technology (CALT), is China's largest entity involved in the development and manufacturing of space launch vehicles and related ballistic missile systems. This premier Academy handles China's entire inventory of liquid fuelled ballistic missiles. It also plays a lead role in China's manned space program through its institutes which specialize in guidance, navigation, and control sub-systems, re-entry vehicles, and launchers. This academy is also working on hypersonic cruise vehicles that operate in the realm of near space in a sub-orbital trajectory, rather than adopting a traditional ballistic flight path. This will be a significant feature of the new generation long-range precision strike vehicles being developed by China.

- **Academy of Aerospace Solid Propulsion Technology (CASC Fourth Academy).** This Academy is responsible for development and manufacturing of solid rocket motors with diameters of 2 m or more.

- **Academy of Space Propellant Technology (CASC Sixth Academy).** This Academy is responsible for research, development, and production of liquid fueled propulsion systems. The latest products of this academy are YF-77 and YF-100, currently China's most powerful liquid oxygen and kerosene rocket engines.

- **Shanghai Academy of Space Technology (CASC Eighth Academy).** This Academy is responsible for designs, develops, and manufactures specialized launch vehicles, satellites, and other aerospace systems prominent among them are weather, synthetic aperture radar, and electronic reconnaissance satellites.

- **China Academy of Space Technology (Fifth Academy).** This Academy is responsible for satellite design, development, and manufacturing. It designs, researches and develops satellite attitude and orbit control systems, including jet propulsion and various guidance, navigation and control sub-systems.

Apart from these supporting academies, the **China Great Wall International Corporation (CGWIC)** under CASC as the principle marketing arm for space launch services. It handles export management and is the international contracting agency which is responsible for the Chinese space commerce. It supplies international customers with satellites, launch services, and other services.[19]The commercial drive of the Chinese space programme can be gauged from the fact that CGWIC's stated business goal is to capture 10% of the international commercial satellite market and 15% of the launch market by 2015.[20]

China Aerospace Science & Industry Corporation (CASIC)

This is the second major industrial establishment for Chinese space-related R&D and production, which specializes in conventional defence and aerospace systems like tactical ballistic missiles, anti-ship and land attack cruise missiles, air defence missile systems, direct ascent anti-satellite interceptors, operationally responsive tactical micro satellites, and associated tactical satellite launch vehicles. The organization chart of CASIC is highlighted below:

---

19 "Launch Services Management," CGWIC Website, at http://cn.cgwic.com/ LaunchServices/index.html on 10 March 2011.

20 XinDingding, "China Seeks Bigger Share Of Global Satellite Market," *China Daily*, October 21, 2010 at http://www.chinadaily.com.cn/china/2010-10/21/content_11437002. htm.

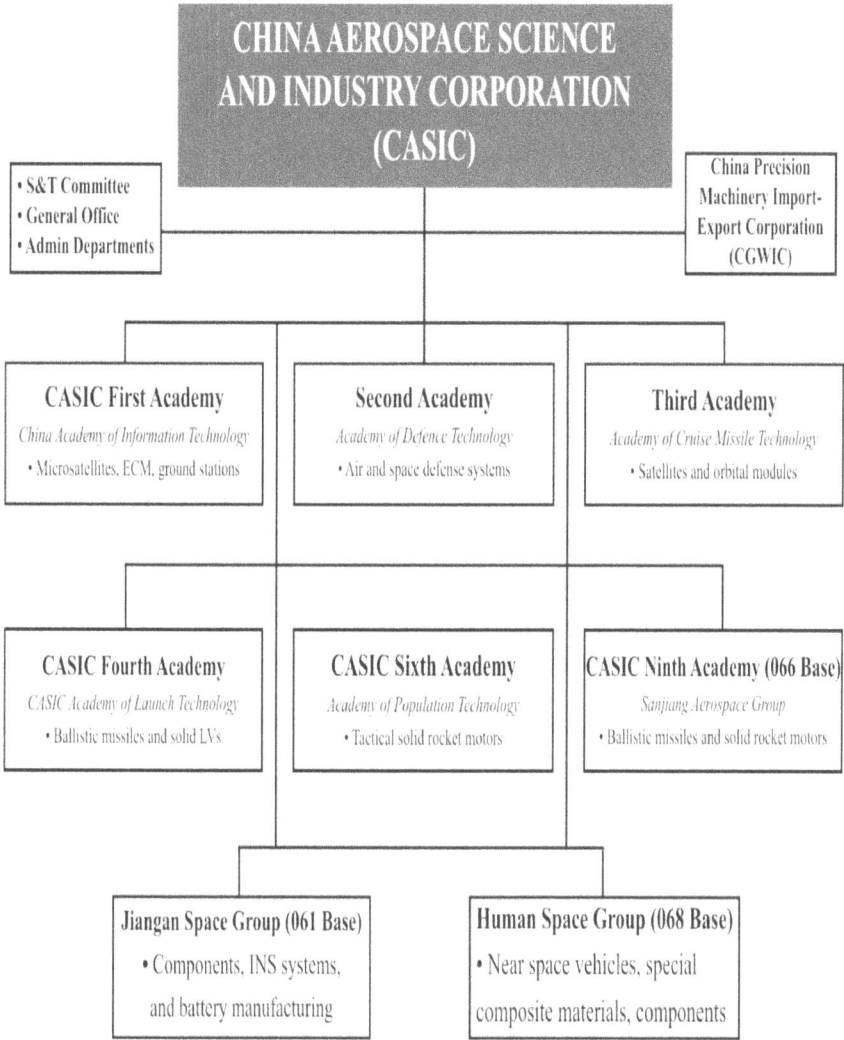

CHINA AEROSPACE SCIENCE AND INDUSTRY CORPORATION (CASIC)

- S&T Committee
- General Office
- Admin Departments

China Precision Machinery Import-Export Corporation (CGWIC)

CASIC First Academy
China Academy of Information Technology
- Microsatellites, ECM, ground stations

Second Academy
Academy of Defence Technology
- Air and space defense systems

Third Academy
Academy of Cruise Missile Technology
- Satellites and orbital modules

CASIC Fourth Academy
CASIC Academy of Launch Technology
- Ballistic missiles and solid LVs

CASIC Sixth Academy
Academy of Population Technology
- Tactical solid rocket motors

CASIC Ninth Academy (066 Base)
Sanjiang Aerospace Group
- Ballistic missiles and solid rocket motors

Jiangan Space Group (061 Base)
- Components, INS systems, and battery manufacturing

Human Space Group (068 Base)
- Near space vehicles, special composite materials, components

Organization Chart of CASIC [21]

Like CASC, even the CASIC functions through different academies which specialize in a particular weapon / space system. The details are as under:-

---

21 "China's Evolving Space Capabilities : Implications for US Interests" a report by Mark A Stokes and Dean Cheng

- **Academy of Information Technology (CASIC First Academy).** This academy deals with design and manufacturing of operationally responsive tactical micro satellites that could be launched on solid-fueled launch vehicles. It also is engaged in R&D satellite applications and GPS/inertial guidance units. Its most prominent product is the Hangtian-Tsinghua-1 (HT-1) 50 kg micro satellite that operates in a sun synchronous orbit. This academy also specializes in space-based and missile-borne electronic countermeasure (ECM) research and development.[22]

- **CASIC Second Academy.** This academy deals with kinetic kill counter-space systems, and happens to be China's largest producer of air defence missile systems. The prominent defence products include the Hongqi-series of surface to air missile systems, including the missile, radar, and associated ground equipment. This academy also designed, developed, and produced the space intercept systems that were tested in January 2007 and January 2010.[23]

- **CASIC Third Academy.** This academy deals with design, development and production of cruise missiles, other aerodynamic vehicles and propulsion systems, and associated launchers. It is also believed to be engaged in R&D on air-breathing supersonic combustion ramjet (scramjet) technology in support of a national hypersonic cruise vehicle program.

- **CASIC Fourth Academy.** This academy deals with design, development and manufacturing of the DF-21 MRBM and associated variants.

- **CASIC Six Academy.** This academy deals with solid motor used for operationally responsive satellite launch vehicles. It manages smaller diameter motors, including

---

22 Formed in 1978 in Nanjing, the 8511 Institute is the aerospace industry's main electronic and infrared countermeasures entity. It manages an integrated test and manufacturing facility in Nanjing's Jiangning Science Park.

23 Chinese Military Network, in China.com, January 13, 2010, at http://military.china. com/zh_cn/critical3/27/ 20100113/15774945. html.

kick motors designed to boost communications satellites to geosynchronous orbit.

- **CASIC Ninth Academy.** This academy deals with cruise missile and its sub systems. Its most prominent product is the DF-11 short range ballistic missile.

- **Jiangnan Aerospace Group.** This academy is a primary supplier of specialized missile components and software. Its 20 institutes and factories develop and produce missile-related guidance, navigation, and control software, composite materials, and a range of components, including aerospace-qualified fasteners, gyroscopes, autopilot systems, batteries, micro-motors, and fuel gauges.

- **Hunan Space Bureau (068 Base).**This academy deals with special materials and components, such as magnets, diamond coatings, and antennas. More recently, the base has become a key center for R&D and production of reconnaissance platforms operating in near space.

As regards the business with foreign entities and countries, CASIC's main agency handling export management is the **China Precision Machinery Import-Export Company (CPMIEC).**[24] However, most of the CASIC academies and institutes appear to conduct independent international business transactions.

Chinese Space Programme

China has established itself as a leading space power, wherein its manned space program has become a vehicle for pushing the limits of human innovation to create new technologies with diverse applications. Since manned spaceflights are an internationally recognized symbol of progress and wealth, China having achieved this feat has thus achieved international prestige and national pride. Besides the national prestige, national resources dedicated toward manned space programs by China also have strategic implications which may be leveraged to enhance its militarily capabilities.[25]

---

24 For background on CASIC, see http://www.casic.com.cn/n16/index.html.

25 The Manned Space Program Office website at http://www.cmse.gov.cn/AboutUs/list. php?catid=9.

In the past several years, China's space efforts have become increasingly prominent. Recent Chinese achievements have included the third manned *Shenzhou* mission and a spacewalk, expansion of the indigenous Chinese Beidou (Compass) satellite navigation system, launch of Data Link satellites for deep space exploration and deployment of a range of new remote sensing satellites, such as the *Yaogan* series. At the same time, there has been growing concern amongst the other space powers that China may be pursuing a policy of space dominance, including programs specifically oriented toward counter-space operations. The best known example supporting this claim is the 2007 anti-satellite (ASAT) test, which not only generated enormous orbital debris, but also debates about the Chinese intentions and its effort towards weaponisation of space. Since then, the Chinese have conducted further tests with potential anti-satellite implications. In January 2010, they undertook a test that involved "two geographically separated missile launch events with an exo-atmospheric collision."[26]Between June and August 2010, two Chinese satellites, SJ-06F and SJ-12, were engaged in orbital rendezvous maneuvers that appeared to include "bumping" into each other.[27]Such activities, undertaken deliberately, would be useful for practicing docking maneuvers or anti-satellite operations. In addition, contrary to international custom, China gave no prior notice of any of these tests, which has heightened concerns and added to the opaqueness of the China's space program.

Shenzhou Manned Space Program

This program called Project 921 is China's largest space program in terms of scope and participation by defence industries. CASC Fifth Academy plays a leading role in design, development, and manufacturing of orbital modules. The program management of the manned space program was centralized within the GAD's China Manned Space Engineering Office. The origins of China's current manned space program can be traced to the late 1980s when Chinese

---

26 Agence France-Presse, "China Did Not Notify US Before Anti-Missile Test," Google News, January 12, 2010, at *http://www.google.com/hostednews/afp/article/ALeqM5gIyJwTWQjzwLtHke9NhVHNS7qiHQ* (July 15, 2011).

27 Brian Weeden, "Dancing in the Dark: The Orbital Rendezvou s of SJ-12 and SJ-06F," *The Space Review*, August 30, 2010, at *http://www.thespacereview.com/article/1689/1* (July 15, 2011).

leaders convened a series of conferences and eventually decided to develop a space capsule, rather than a space plane or space shuttle design, which eventually became the Shenzhou spacecraft, which means 'Divine Craft or Divine Vessel'. The core area of each Shenzhou space craft and its achievement is as under

- Shenzhou-1 launch in 1999, tested the space capsule's control systems and interfaces.

- Shenzhou-2 focused on the capsule's environmental systems.

- Shenzhou-3 mission in March 2002 expanded environmental testing, and remained in orbit for six months during which testing of various payloads took place.

- Shenzhou-4 certified the capsule's emergency rescue systems.

- Shenzhou-5 brought the most momentous event in the history of China's space program took place in October 2003, when China sent then-Lieutenant Colonel Yang Liwei of the PLA Air Force (PLAAF), into space on board Shenzhou-5.[28] It orbited the earth 15 times before landing safely in Inner Mongolia.

- Shenzhou-6 was used as the second manned spaceflight to establish a foundation for docking a space laboratory, with the ultimate objective of maintaining a larger space station for long-term scientific experiments.

- Shenzhou-7 was a part of China's second phase of manned space flights. This space ship brought in a moment of pride for the Chinese space programme when its taikonaut carried out the first spacewalk as part of the program's mission in 2008. In this historic flight, Chinese astronaut Zhai Zhigang emerged from the Shenzhou-7 capsule orbiting the earth and waived a Chinese flag triumphantly. His 20-minute stay in outer space was witnessed by millions of Chinese on the earth through live broadcast recorded by his fellow

---

28 BBC News on 15 Oct 2003, "China puts its first man in space", accessed through http://news.bbc.co.uk/2/hi/asia-pacific/3192330.stm

taikonauts from inside the Shenzhou -7 craft on 28 Sep 2008.[29] The exercise is seen as key to China's ambition to build an orbiting station in the next few years.

- Shenzhou-8 was launched on 01 Nov 2011 and helped the Chinese taikonauts to conduct another milestone in space programme, wherein they successfully carried out their first automated docking mission with their Space module Tiangong-1(The Heavenly Palace).

- Shenzhou-9 created another history when it became the instrument of China's first manned docking mission on 16 Jun 2012. This was a significant achievement for the Chinese space programme, which nurtures ambitions of prolonged stay in space and deep space missions. [30] This mission also brought fame to Chinese space programme as it carried the first lady, Maj Liu Yang to space on 16 Jun 2002 and broke the gender bias in space exploration. [31]

- Shenzhou-10 was launched on 11 Jun 2013 to re-validate its docking capabilities and also to perform longest stay of 15 days in space module Tiangong-I for space experiments. This was the end of Phase –I of the manned space programme, which will be followed by launch of Tiangong-2 by 2015 for future manned space module, a precursor to the Chinese Space Station, which they plan to establish by 2020.[32]

Lunar Exploration Program

China started its lunar exploration program in 1995 under 863 Program. One of the motives of the Chinese Lunar Mission was to launch its orbiters (Change) to moon for exploring the prospects for

29 BBC News on 27 Sep 2008, "Chinese Astronaut Walks in Space", accessed through http://news.bbc.co.uk/2/hi/science/nature/7637818.stm

30 Reported by Xinhua News in http://www.xinhuanet.com/english/special/shenzhou9/index.htm

31 BBC News on 16 Jun 2002, "Profile of Liu Yang, China's First Woman Astronaut", accessed through http://www.bbc.co.uk/news/science-environment-18471236

32 Xinhua News Agency quoted by web portal Space Daily on 27 Jun 2013 "China Plans to Launch Tiangong-2 Space Lab around 2015", accessed through http://www.spacedaily.com/reports/China_plans_to_launch_Tiangong_2_space_lab_around_2015_999.html

mining lunar Helium-3 as a replacement for fossil fuels. The lunar program was structured into three phases.[33]

- **Phase I.** It was designed to be a demonstration of its technological prowess, involving the launch of lunar orbiters Change-1 in 2007 and Change-2 in 2010. The mission of Change-1 was creation of a 3D map of the lunar surface, studying the presence and distribution of useful elements and minerals, and measuring solar winds and their impact on the earth and moon. The Change-1 spacecraft was launched on 24 Oct 2007 and orbited the moon for approximately one year.[34]

    China further launched Change-2 on 01 Oct 2010 as a part of its Orbiter Lunar Mission to focus on future lunar rover landing zones, surveying of Lagrangian points, and testing of upgrades to the Chinese space tracking and control network. Change-1 and Change-2 were to obtain precise images and maps of the lunar surface and analyze the content and distribution of useful elements and minerals, such as uranium and titanium. Once in orbit around the moon, the spacecraft took pictures of the lunar surface with a stereo camera/spectrometer imager, including photographs of the northern and southern pole regions; investigated and analysed elements on the lunar surface for locations of large deposits of elements; and took measurements of the lunar subsurface, including measurements of Helium-3.

- **Phase II.** Under this lunar exploration program "Soft Lander" crafts will be launched to moon's surface. This programme is scheduled to commence in end 2013 with launch of Change-3[35], which will be followed with Change-4

33 "China's Space Program: Civilian, Commercial, and Military Aspects," <http://www.highfrontier.org/Archive/hf/Finkelstein%20China's%20Space%20Program.pdf>.

34 OuyangZiyuan, "The Scientific Objectives of Chinese Lunar Exploration Project, presentation before the 9th International Symposium on Physical Measurements and Signature in Remote Sensing (ISPMSRS), Beijing, China, October 17-19, 2005, at http://www.geog.umd.edu/ispmsrs2005/OuyangZiyuanAbs.htm.

35 Leonard David in Space News, 23 Jun 2013, "China Readying I st Moon Rover for Launch this Year", accessed through http://news.yahoo.com/china-readying-1st-moon-rover-launch-193901386.html

in 2014. Change-3 will become China's first craft to attempt a soft landing and rover deployment on the surface of the moon. The full mission profile is believed to include docking, controlling, and mapping missions. Two remote controlled rovers are to be deployed to conduct surface investigations.

- **Phase III.** It is slated for 2017 with the launch of Change-5 on the LM-5E heavy launch vehicle for collecting samples from the lunar surface. The program has plans to send a manned lunar landing mission which is slated after 2025.[36]

China's Space Station Program

China launched its first space module Tiangong-1 or the "heavenly palace," on 29 Sep 2011. This module was an experimental test bed for the Chinese to conduct their 'orbital rendezvous and docking capabilities. The Chinese have planned to place larger habitable modules by 2020 to enable prolonged stay of Chinese taikonauts. Tiangong-1 is planned to be de-orbited in 2013, after the successful docking with Shenzgou-10 mission which was the last part of this phase. It will be replaced by Tiangong-2 space module by 2015, which will eventually pave the way for the ISS equivalent by China by 2020.

Tiangong-1 consists of a main module with two detachable modules on either side. The "Core Cabin Module" weighs 8.5 tons and is equipped with two docking ports, and will have a two-year lifetime in earth's orbit. This unpiloted spacecraft was the basis for tests on docking technologies, wherein Shenzhou-8, an unmanned spacecraft created the first space docking in Nov 2011 and was followed by Shenzhou -9 on 16 Jun 2012 [37] and Shenzhou-10 on 11 Jun 2013 for re-validating the manual and automated docking experiments.

The Tiangong-2 space module called "Laboratory Cabin Module," is scheduled to be launched by 2015, which is designed to

---

36 Bruce Sterling, "Chinese Manned Moon Landing, 2025," 19 September 2010, at http://www.wired.com/beyond_the_beyond/2010/09/chinese-manned-moon-landing-2025/>

37 Clara Moscowicz, "China Shifts Space Station Project into Overdrive," *Space.com*, 15 April 2010 at http://www.space.com/8224-china-shifts-space-station-project-overdrive.html.

test technologies for larger space stations with capabilities to sustain long-term living conditions for astronauts. The Tiangong-2 will be able to accommodate three astronauts for about 20-40 days at a time. This module will be utilized to study regenerative life-support technologies and methods for replenishing fuel and air during missions. The experience gained from the Tiangong programme will help the Chinese to establish a technology base to launch a fully functioning space station by 2020.[38]

Once completed, the station will consist of an 18.1-meter-long center module and two 14.4-meter laboratory modules, plus a manned spaceship and a cargo craft. The entire space station will weigh around 60 tons. Astronauts will eventually take up residence on the station to conduct research in the fields of astronomy, microgravity, and biology. China's primary motivation for investing in a space station could be to support its lunar program. Chinese officials have stated that the country's space technology will be compatible with that used in the International Space Station, so that modules from other countries will be able to dock with its station.[39]

MILITARY APPLICATIONS OF CHINESE SPACE PROGRAMME

Space Based Intelligence, Surveillance and Reconnaissance

The presence of the US forces and ships in the Asia-Pacific and the South/East China sea compelled the Chinese leadership to explore the possibilities of reliable surveillance over US military assets transiting or stationed in China's neighbourhood. Space-based surveillance systems appeared to be the best way for a non-intrusive surveillance over the US forces. The space based assets have enhanced China's ability to conduct military operations away from its shore. The space-based sensors are now capable of providing images to PLA which are necessary for mission planning

---

38 Clay Dillow, "China Announces Plan to Build a Manned Space Station of its own Within Ten Years," *Popular Science*, 26 April 2011, 27 June 2011 <http://www.popsci.com/technology/article/2011-04/china-plans-build-manned-space-station-its-own-within-ten-years>.

39 Clara Moscowicz, "China's Lofty Goals: Space Station, Moon, and Mars Exploration," *Space.com*, 6 Dec 2010, 27 June 2011 <http://www.space.com/166-china-lofty-goals-space-station-moon-mars-exploration.html>.

and counter offensive against any threat to China's national interest. From this perspective, China has fielded electro-optical radar and other space-based sensor platforms that can transmit images of the earth's surface to ground stations in near-real time. Coupled with high end surveillance capabilities, new generation secured satellite communications offer China a reliable means of communication to the deployed Chinese forces.

In order to have a reliable strike capability, China will now use its high resolution, dual-use space-based synthetic aperture radar (SAR), electro-optical (EO), and possibly electronic intelligence (ELINT) satellites for surveillance and targeting. China's space industry is reportedly nearing completion of its second generation SAR satellite, and its EO capabilities have been progressing steadily. Existing and future data relay satellites and other beyond line of sight communications systems could relay targeting data to and from the theater and/or Second Artillery's operational-level command center.[40]

China is now working on EO satellites which will have digital camera technology, as well as space-based radar for all-weather, 24 hour coverage, so as to monitor radar and radio transmissions for the purpose of ELINT. China is also deploying robust weather satellites, oceanography satellites, specialized satellites for survey and mapping, and possibly space-based sensors capable of providing early warning of ballistic missile launches. The details are discussed below:

- **Electro-Optical Satellites.**

China launched its first experimental imagery system November 1975, which could only be made operational in September 1987 after the Fanhuishi Weixing-1 (FSW-1) or recoverable satellite was launched from Jiuquan Space Launch Center. The FSW-1 provided wide area imaging and has the capability to orbit for 8 days. This series was replaced by a more advanced FSW-2 satellite, which had maneuvering capability and could carry 2,000 m of film and had a resolution of 10 meters. The FSW-2 used to orbit for 15 or 16 days before returning to earth.

---

40 "China Blasts Off First Data Relay Satellite," *Xinhua News Agency*, April 26, 2008.

In the 1980s, China collaborated with Brazil for development of a more capable electro-optical imaging satellite as China-Brazil Earth Resources Satellite (CBERS) program. In Oct 1999, China was able to successfully launch ZY-1 (CBERS-1) which provided 20 m resolution images with near-real time surveillance capability that could be transmitted digitally to a ground station. After two more ZY-1/CBERS satellites in 2003 and 2007, China went on to launch its second generation ZY-2 satellites, which were apparently in a lower orbit, offering finer resolution images with the same sensor array. The ZY-2 satellite launched in 2000, 2002, and 2004, are said to have military reconnaissance capabilities.[41]

- **Synthetic Aperture Radar Satellites.**

China's first indigenous Earth Observation satellite was a dedicated military SAR satellite Yaogan-1, which was launched from Taiyuan Satellite Launch Center on 27 Apr 2006. This was followed by Yaogan-2, which was a reconnaissance satellite, developed and manufactured by CASC Fifth Academy, and launched on May 25, 2007.[42] Equipped with either optical or synthetic aperture radar (SAR) sensors, many more Yaogan-2 series satellites were launched by China to provide the much needed surveillance capabilities required by PLA and also the capabilities of disaster monitoring.

The military application of the Yaogan series satellites can be appreciated from the fact that, these SAR equipped satellites are a core component of militarily-relevant surveillance architecture supporting over-the-horizon (OTH) targeting of surface assets. They use a microwave transmission to create an image of maritime and ground based targets and can operate 24X7 in all weather conditions, and is therefore well suited for detection of ships in a wide area, a capability which China considers essential to secure its national interest. The SAR imagery plays a crucial role in recognition of ships at sea and is

---

41 "National Remote Sensing Center, Ministry of Science and Technology website, undated, at http://www.most.gov.cn/zzjg/zzjgzs/zzjgsyygzx/index.htm.

42 "Successful Launch of the Yaogan-2", Xinhua, May 25, 2007, at http://news.xinhuanet.com/politics/2007-05/25/content_6152111.htm.

also able to detect ship wakes from which information on ship speed and heading can be analysed.[43]

- **Electronic Reconnaissance Satellites.**

China considers electronic reconnaissance a necessity to accurately track and target the US carrier strike groups operating in the Asia Pacific, in near real time. This capability can support China's long-range precision strike capability, including its anti-ship ballistic missile (ASBM) system against the perceived maritime threat.

The important electronic reconnaissance satellites for China are the Shijian-6 and Yaogan-9 satellites. The Shijian-6A and Shijian-6B satellites were launched in tandem in September 2004.[44] Based on the expertise gained from this dual satellite project, China has so far launched a number of Shijian satellites. The Shijian-12, launched on 15 Jun 2010, was considered significant for special experiment, wherein it was maneuvered near the SJ-06F between 20 Jun 10 and 16 Aug 10 for a 'co-orbital rendezvous' perhaps for an inspection mission.[45] However, space analysts feel that this was a preparatory phase for China's co-orbital ASAT test.

- **Oceanographic Satellites**

These satellites are dual use satellites, which apart from special naval operations, are also useful in disaster warning, recovery, and response, support for fishing, exploitation of maritime resources. The multispectral sensors fitted on these satellites can easily detect ships at sea. China achieved this capability and launched Haiyang-1A and 1B satellites in May 2002 and April 2007 respectively. The satellites, integrating electro-optical and other sensors, are mainly used for monitoring water

---

43 Chen Deyuan and TuGuofang, "SAR Image Enhancement Using Multi-scale Products for Targets Detection, Remote Sensing Journal [*Yaoganxuebao]*), March 2007, pp. 185-192.

44 "Shijian-6 Tandem Twin Launch Successful", *China Spacesat Company*, October 8, 2010, at http://www.spacesat.com.cn/newdfh/xwzx/ArticleShow.asp?ArticleID=260.

45 Brian Weeden, " Dancing in the Dark: The Orbital Rendezvous of SJ-12 and SJ-06F, *The Space Review*, August 30, 2010, at http://www.thespacereview.com/article/1689/1.

CHINA'S RESURGENCE IN SPACE

color, water environment and temperatures.' These satellites were succeeded by HY-2 in 2009, which integrates microwave technology to detect sea surface wind field, sea surface height and sea surface temperature. Research and development on a more advanced ocean monitoring system incorporating SAR technology, the HY-3, is well underway specifically for the maritime surveillance. HY-3 integrates multiple sensors, such as a multi-spectral imager, synthetic aperture radar, microwave scatterometer, radiometer and radar altimeter. [46] It is expected that China is planning to upgrade its maritime satellite network by 2020 to secure its maritime environment from any and every kind of threat to its national interest and maritime security.

- **Meteorological Satellites**

China's weather satellite programme started in 1988 with Fengyun (FY-1A), making it third nation to launch its own meteorological satellites. Since then, the PRC has launched four FY-1 weather satellites into polar orbit, five FY-2 geosynchronous weather satellites, and two FY-3 satellites that were boosted into polar orbits. The FY-3 weather satellite, with almost a dozen all weather sensors, is China's most advanced space asset providing meteorological support to PLA and other users. [47]

The next generation geosynchronous weather satellite FY-4 is expected to enter service in 2014, will be a dual use asset for use by both the PLA as well as China Meteorological Bureau.[48]These satellites will collect and provide strategic weather reconnaissance data for civilian and military purposes. An accurate assessment of current and future weather conditions, such as cover, atmospheric moisture, winds, temperature, and ocean currents, is critical for a range of military operations.

---

46 Jiang Xingwei, Lin Mingsen, and Tang Junwu, "The Programs of China Ocean Observation Satellites and Applications," National Satellite Ocean Application Service briefing, February 26, 2008

47 "Delivery of Four FY-3 Payloads and Satellite Testing",*Sina News*, August 8, 2006, accessed January 20, 2012, http://tech.sina.com.cn/d/2006-08-08/15291076129.shtml.

48 The First Satellite in Fengyun-3 Series to Be Launched in 2006", *Xinhua*, March 23, 2004, at http://news.qq.com/a/20040323/000009.htm.

Weather satellites can measure electromagnetic conditions in the ionosphere that could affect OTH radar and communication systems. They also can provide militarily useful data associated with complex maritime environments and terrains, including observation of targets under camouflage or perhaps even underground.

Space-Based Communications and Navigation Satellite Systems

Space-based systems are the best means for communicating over the horizon. This ensures a Network Centric Warfare capability by equipping sailors, airmen, and soldiers with a common operational picture that significantly increases their situational awareness of the battlefield. The details of the Chinese satellites used for ensuring network centric operations are discussed below:

- **Communications Satellites.**

Ever since the launch of its first communication satellite in 1984, the basic intent of the Chinese leadership was to raise the level of education in China. With time, China's communications satellite capacity has grown in sophistication. However, with the successes of DFH-4 telecommunications satellite, the PLA appears to have invested in dedicated military communication satellites. Fenghuo-1 also known as ChinaSat-22, which was launched in Jan 2000 became PLA's first dedicated military communications satellite. With development in technology, these satellites were equipped with steerable spot beams operating in the Ku Band.[49]

- **Data Relay Satellites.**

With increase in the number of satellites launched by China, there was a need to integrate linkages between the satellites for data transfer. This was essential to facilitate Chinese deep space missions. Tianlian-1launched in April 2008 was the first generation data relay satellite of China which was followed by second in July 2011. These satellites allow space sensors

49 http://www.cgwic.com/In-OrbitDelivery/CommunicationsSatellite/DFH-4Bus.html.

to operate beyond line of sight of ground stations.[50]Chinese launched their third tracking and data relay satellite Tianlian-1C on 25 Jul 12 to complete their satellite data relay system. This satellite will be used for tracking and data relaying on manned missions and to other Chinese satellites (both civil and military). Once in the designated position, this satellite will allow almost non-interrupted communication during a manned mission.[51]

- **Navigation Satellites.**

China wanted to have an independent navigation system of its own rather than depending upon the US GPS system, which could be switched off or manipulated by the US in times of crisis. China's first generation navigation satellite system, the Beidou-1 was launched with first satellite, Beidou-1A on 30 Oct 2000, followed by Beidou-1B on 20 Dec 2000 and third satellite Beidou-1C was put into orbit on 25 May 2003. It is an experimental regional navigation system, which consists of four satellites (three working satellites and one backup satellite). Unlike the American GPS, Russian GLONASS, and European Galileo systems, which use medium Earth orbit satellites, Beidou-1 uses satellites in geostationary orbit. This means that the system does not require a large constellation of satellites, but it also limits the coverage to areas on Earth where the satellites are visible. Beidou system started its regional open service in the Asia Pacific region with an accuracy of 10 meters, timing of 0.2 microseconds, and speed of 0.2 meters/second. Apart from China, as of now Laos, Thailand, Brunei are using this system and Pakistan has signed the MoU to use Beidou system rather than the US GPS.

In Phase II of the project, the Beidou 1 system will be replaced by Beidou-2 which will have a constellation of 35 satellites completely covering the earth. Beidou-2 will include 5 geostationary orbit satellites and 30 non-geostationary satellites (27 in medium earth orbit and 3 in inclined geosynchronous

50 "China Launches New Data Relay Satellite," Xinhua Net, July 12, 2011, at http://news.xinhuanet.com/english2010/sci/2011-07/12/c_13978690.htm

51 News report in Orbiter Forum on 26 Jul 12 titled "Tianlian-1C, Long March 3C, July 25, 2012" accessed through http://www.orbiter-forum.com/showthread.php?t=28469

orbit). There will be two levels of services provided; free service to civilians and licensed service to Chinese government and military users. Though the free civilian service will have a 10 meter location-tracking accuracy, will synchronize clocks with an accuracy of 10 ns, and measure speeds within 0.2 m/s. But the restricted military service will have an accuracy of 10cm. To date, this service has been granted only to the People's Liberation Army and to the Military of Pakistan.Beidou-2 is likely to provide global coverage by 2020.

Micro-satellite Programs

China appreciates that their space programme is vulnerable to the US space operations. Hence to cater for a crisis situation, it has put plans to augment its existing space-based assets with micro satellites launched on solid-fueled launch vehicles "on demand". These micro satellites weighing between 10 and 100 kg are cost effective and operationally responsive space capability for a specific mission or to temporarily support an important constellation. They also serve as experimental technology test beds and have also served as technology demonstrator for counter-space operations, including ASAT kinetic kill vehicles. [52]

China's initial technology demonstration programs for micro satellite included Tsinghua-1 satellite which was developed in conjunction with the UK's University of Surrey and launched in June 2000. Weighing just 50 kg, this micro satellite conducted experiments on satellite-borne navigation, multispectral remote sensing, and store and dump downlink communications. Another micro satellite tested by China was Naxing-1, which was launched on 18 Apr 2004. It rode piggyback on the Shiyan-1as a test bed for an on-board miniature inertial measurement unit and digital imagery.[53]

China's independent micro satellite is the Pixing-1, developed

---

52 China Space Network, May 6, 2011, at http://www.space.cetin.net.cn/index. asp?modelname=new_space%2Fnews_nr&FractionNo=&titleno=XWEN0000&rec no=75869.

53 "The First Indigenously Made Nano-1 Satellite Has Been Launched Successfully",S&T Daily,, April 18, 2004, at http://www.stdaily.com/oldweb/gb/kjzg/2004-09/28/ content_305818.htm

by Zhejiang University for defence-related basic research. It was launched in conjunction with the Yaogan-2 on 25 May 2007. This was followed by two more Pixing micro satellites as piggyback payloads on the Yaogan-11 in 22 Sep 2010 as test platform for digital imagery, data storage and management; downlink communications, attitude control; MEMS inertial measurement unit, thermal control, and other missions.[54]

Another significant micro satellite the BX-1, a very small cube approximately measuring 40 centimeters each side and weighing around 40 kilograms was launched   aboard Shenzhou-7 on 25 Sep 2008. China declared that the purpose of the BX-1 was to provide images of the Shenzhou-7 space ship and to demonstrate the ability to inspect the orbital module and conduct some limited proximity operations. It also carried out a data relay experiment. However, some observers have concluded that the BX-1 was actually a test of some of the capabilities required for a co-orbital anti-satellite (ASAT) attack.[55] Chuangxin-1 (CX-1) was another micro-satellite weighing 88 Kg, launched by China in Oct 2003, which was a prototype low earth orbit telecommunication satellite. Similarly Beijing-1 designed and manufactured by the Surrey Satellite Technology Ltd (SSTL) for the Disaster Monitoring was launched in Oct 2005.[56]

Apart from the micro satellites, China also developed operationally responsive solid fuelled rockets Kaituozhe-1 (KT-1) for placing micro satellites into orbit 'on demand'. It is capable of delivering a 50 kg payload to a 400 km altitude sun-synchronized orbit. This category of small launch rockets were to serve the domestic and foreign market for boosting small and micro satellites with weights less that 100 kg into low earth orbit. The advance version KT-2 is said to be a three or four-stage launch vehicle for geosynchronous and polar orbits missions with an estimated payload

54 "Nation's First Kilogram-Level Micro-Satellite Operating Smoothly for Eight Days and Nights", Xinhua Network, September 30, 2010, at http://news.sina.com.cn/c/2010-09-30/222621204836.shtml.

55 Brian Weeden in "China's BX-1 microsatellite: a litmus test for space weaponization" published in The Space Review on 20 Oct 2008

56 Peter B. de Selding, "Surrey To Build Three Optical Imaging Satellites for Chinese Firm," Space News, June 29, 2011, at http://www.spacenews.com/contracts/110629chinese-firm-orders-three-optical-imaging-satellites-from-surrey.html.

capability of 300 kg. [57]

Organisational Structure of China's Counter Space Operations

China views space as a national priority. Accordingly, it plans for effective utilization of the space domain and simultaneously develops capabilities to deny others the use of space. This is central to their defence modernization goals. Over the next 10-15 years, the PLA is expected to expand its integrated architecture of sensors, communications, long-range precision strike assets, and other joint operational capabilities. This will enable China to operate freely in space and deny others freedom of operation in space, if required.

The **General Staff Department** (GSD) provides for an organizational infrastructure for developing requirements in space and operating the ground segment to support space operations. GSD, GAD, and Service missions comprising of armed forces will together integrate and synergise their working and will eventually create China's "National Aerospace Security System". [58]

The functions assigned to various departments under the General Staff Department are discussed below:

- **First Department (Operations Department)**

  It develops requirements for and manages joint military use of navigation, geodetic, metrological, and oceanographic space systems. The Survey and Mapping Bureau manages the ground segment of the Beidou satellite positioning system and is also believed to operate a very long baseline interferometer (VLBI) network of radio telescopes that support China's space tracking system. The Weather and Hydrological Bureau manages military meteorological satellite data and also oversees a specialized unit responsible for space weather analysis and forecasting.

---

57 Gu Ti, "Kaituozhe: New Choice for Small Satellite Launches," *Aerospace China*, November 2002, p. 2.

58 "Establishing a National Aerospace Security System", *Science News,* February 24, 2002, at http://www.cas.cn/xw/zjsd/200202/t20020224_1683499.shtml

- **Second Department (Intelligence Department)**

  It works towards the development of space-based reconnaissance operational requirements and operation of ground receiving stations. The GSD Space Reconnaissance Bureau appears to be primarily focused on electro-optical (EO) and synthetic aperture radar (SAR) remote sensing operations.

- **Third Department (Technical Reconnaissance Department)**

  It functions as China's primary signals intelligence (SIGINT) collection and analysis unit. It manages collection, translation, and analysis of communications intelligence (COMINT). It operates satellite monitoring facilities and through its units, intercepts satellite communications from sites throughout China and possibly from space-based sensors.

- **Informatisation Department**

  It develops operational requirements for PLA and oversees use of dedicated military communications satellites, such as the Fenghuo and Shentong systems (ChinaSat-22 and ChinaSat-20).

- **Strategic Planning Department**

  It is responsible for organizational transformation, strategic resource allocation, and departmental and "domain" coordination between departments like GSD and GAD. This department may play a central role in force planning for future space operations. It supports the Chief of General Staff in assessing threats to National security from space.

- **Fourth Department (Radar and Electronic Countermeasures Department)**

  It is responsible for radar and electronic countermeasures and appears capable of disrupting adversary's communications, navigation, synthetic aperture radar and other satellites system. It specialies in satellite jamming and can also operate electronic reconnaissance satellite ground receiving stations to support joint targeting.

The organization chart of General Staff Department is given below:

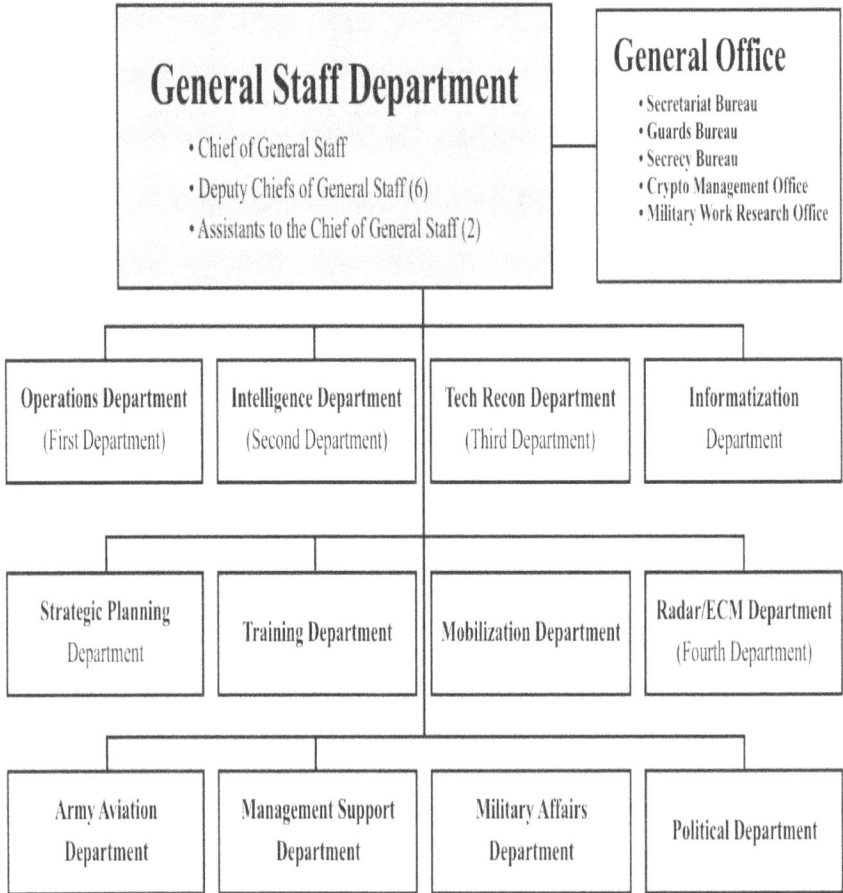

General Staff Department	General Office
• Chief of General Staff • Deputy Chiefs of General Staff (6) • Assistants to the Chief of General Staff (2)	• Secretariat Bureau • Guards Bureau • Secrecy Bureau • Crypto Management Office • Military Work Research Office

Operations Department	Intelligence Department	Tech Recon Department	Informatization
(First Department)	(Second Department)	(Third Department)	Department

Strategic Planning	Training Department	Mobilization Department	Radar/ECM Department
Department			(Fourth Department)

Army Aviation	Management Support	Military Affairs	Political Department
Department	Department	Department	

Organisation Chart of GSD [59]

China's Counter-space Capabilities

China intends to use its space assets to seek asymmetric advantage in any war with the US. Only an asymmetric advantage can give China ability to deny an adversary access to its space assets, offer military advantages in land, air, maritime, and information domains. In view of the overdependence of Space Powers on space assets

---

59 "China's Evolving Space Capabilities : Implications for US Interests" a report by Mark A Stokes and Dean Cheng

and the vulnerabilities in their space infrastructure, China can target their space assets and seek to deny advantages gained through space capabilities.

China has thus developed a range of passive and active counter-space technologies which are discussed below

- **Space Surveillance Network.** China appreciates that any counter-space operations is fully dependent on a secure space surveillance network. Accordingly it is gradually developing the requisite supporting infrastructure which will provide it the ability to track and mitigate space debris and can also fulfill its space surveillance requirement towards military purpose. The PLA and civilian counterparts also have been enhancing national satellite laser range finding capabilities, and investing in radar systems for satellite surveillance and tracking.[60]

- **Kinetic Kill Vehicle Development.** China is working towards creating a "national aerospace security system" to detect and neutralise any threat in/from space.[61] To demonstrate its ASAT capabilities it carried out successful test of a KKV in Jan 2007, amid-course interception in Jan 2010 and a high orbit interception test in thus demonstrating its ability to intercept polar orbiting satellites and rudimentary medium range ballistic missiles during the mid-course of flight. Further to cater for threats in HEO and MEO, China successfully launched a rocket on 13 Jun 2013 from Xichang Satellite Launch Centre, which reached an altitude of 6,200 miles, the highest Chinese sub-orbital flight in 37 years. This rocket is capable of delivering an ASAT to neutralize satellites in MEO and HEO.[62]

---

60 Yousaf Butt, "Effects of Chinese Laser Ranging on Imaging Satellites," Science and Global Security (17), pp. 20–35, 2009, at http://www.princeton.edu/sgs/publications/sgs/archive/17-1-Butt-Effects-of-Chinese.pdf.

61 "Establishing a National Aerospace Security System", *Science News*, February 24, 2002, at http://www.cas.cn/xw/zjsd/200202/t20020224_1683499.shtml

62 NTI Global Security Newswire on 15 May 2013, "China Fires Rocket Capable of Targeting Satellites" , accessed through http://www.nti.org/gsn/article/china-fires-rocket-capable-targeting-satellites-expert/

- *Counter-space Organizations.* The GAD Headquarters Department (CLTC) oversees China's launch centers and satellite tracking and control network. However, in view of the expanding activities in space, PLAAF and Second Artillery have started projecting themselves as competent agencies to handle space operations. The PLAAF proposes to put air and space, both integrated under a single air defence command organization to be managed by them. However, the Second Artillery argues that it should be responsible for all military operations in space. But there is a strong possibility that the aerospace defence domain would be divided along the Karman Line: the PLAAF would assume the air defence mission for threats below 100 km, while the Second Artillery would be responsible for threats above 100 km.[63]

China's Growing Counter-Space Capabilities and the Evolving Space Doctrine

China has emerged as a space power and by the end of this decade it will be formidable space player. Though its space technology may not match that of the US and other space faring nations, yet China's relative advances are significant. The asymmetric advantages of its space systems and capabilities could present challenges to any space power. A survivable space-based sensor architecture which is able to transmit reconnaissance data to ground sites in China in near-real time, facilitates the PLA's ability to project firepower at greater distances and with growing lethality and speed. China is also putting up an ambitious counter-space program, which includes a ground and space based surveillance systems, electronic warfare capabilities, and KKVs. A space surveillance system for detecting and tracking objects in air/space which has low radar cross sections is into finalization and would be deployed soon for an effective and precise counter-space operation. It is therefore evident that China is seeking all these developments in space warfare capabilities so as to ensure an appropriate counter-space capability to counter the US space supremacy and thereby develop asymmetric advantages over the much stronger and modern US forces.

63 Li Guoqiang, "New Strategy of the PLA Air Force", *Wenhuipo*, (Hong Kong), November 26, 2009, at http://paper.wenweipo.com/2009/11/26/PL0911260003.htm

China's military space doctrine is evolving along with its space technologies and capabilities. It has moved towards a doctrine of deterrence in offensive counter-space capability. However, since PLA envisions conflict in space as a distinct possibility, hence it is now seeking a full space war-fighting capability, a step beyond its deterrence posture.

CHAPTER - VI

INDIA'S SPACE AMBITIONS AND POSSIBLE MILITARY EXPLOITS

In 21st century, there would be hardly any industry/institution in the world, which is not using space assets or space enabled services for its growth and development. Satellites have proved to be the **"Engines for National Growth"**, which have enabled sustained growth of the world economy in the 21st century. Today, almost every aspect of human activity and services are dependent on some space based assets or the other. Be it communication, entertainment, navigation, weather forecast, disaster management, rain forecast for cultivation, water zoning, flood monitoring, to name a few. So intrinsically are the daily human activities dependent on satellite enabled services that today, from the time we get up till the time we go back to sleep, life revolves around gadgets and services which are satellite enabled. From switching on TV in the morning for news, weather forecast, share market's trend and analysis, DTH services for entertainment, cellphone usage for connectivity with office and staff, navigating to office and other business establishments using GPS, using internet on the move for latest updates on office matters, swiping credit card for filling up petrol, withdrawing money through ATM, on line transfer of funds for business transactions, reservations for train or airlines, safe transportation through dedicated network of communication and navigation aids, almost anything we can think of, is happening today through satellite enabled services and space assets. Thus commercial utilization of satellites has catapulted countries into phenomenal economic growth; as a result, the demand for satellites is ever increasing. So much so that smaller countries like Sri Lanka, Venezuela, Vietnam, etc have also made sure that they today have their own dedicated satellites. The ones which cannot afford satellites have gone in for hiring of transponders from space faring nations. This shows the economic utility of space based assets in the 21st century and thus a reason for exponential proliferation of satellites. It won't be absurd to say that the basic role of satellites, which was military applications, have now been overtaken by commercial and economic utility for trade and commerce and societal good.

An Overview of Indian Space Security

Space security is a subject which received little or no attention in India until the beginning of the last decade. It has moved up in the priority list of India's security establishment in the aftermath of Kargil War. The event that further shook the Indian leadership from slumber and galvanized their attention on space security in India was China's successful anti-satellite (ASAT) weapons test in 2007. This one event highlighted the threats to India's considerable space assets and infrastructure of almost $12 billion (Rs 60,000 crores). It also brought home to Indian policy planners the necessity and urgency of a long-term planning for a comprehensive space policy. The principal debate on space security in India is now focused on the utility of developing kinetic kill ASAT technologies and the proposed International Code of Conduct for Outer Space Activities.

Indian space program started in 1962 with the sole motive of societal good. Accordingly, space technologies which were developed for use by the people of the country, rather than for any military good. Thus from the early 1960s until the early 2000s, all space activities were directed towards supporting agriculture, education, tele-medicine, tele-education, communications, and remote sensing, which were the primary areas of focus.[1] However, India started to expand its military footprint in space in late 90's. From 2001 onwards, India has launched dual-use satellites that combine civilian and military functions. These include the Technology Experimental Satellite launched in 2001, the Cartosat-2A launched in 2008, and the Cartosat-2B launched in 2010.[2] With proliferation of space technology, the Indian armed forces slowly

1   K. Kasturirangan, "Science and Technology of Imaging from Space," Current Science 87, no. 5 (September 2004): 584–601.

2   Kadayan V. Gopalakrishnan in " Impact of Science and Technology on Warfare" pp 72–73.

started incorporating the elements of space technologies for military use. Like any other modern armed force, Indian armed forces also became increasingly reliant on satellites for their communications, surveillance, and navigational requirements. However, China's successful demonstration of its ASAT capability in 2007 in which it destroyed the FY-1C satellite, brought in a significant change in the strategic thinking in India and a paradigm shift in the matters of "space security." There is some evidence to suggest that Indian government agencies may have begun exploratory efforts to develop anti-satellite weapon technologies in the aftermath of China's ASAT test. However, Indian national security managers at the highest levels continue to maintain that India's uses of space for fulfilling "strategic objectives" are being crafted within the ambit of the 1967 Outer Space Treaty.[3]

Indian Exploits of Space

From a modest beginning of launching rockets from Thumba Equatorial Rocket Launching Station in 1963 in Thiruvanthpuram for collection of meteorological data, to launching the indigenous RISAT-I on 26 Apr 2012, India has come a long way in satellite manufacturing and rocket launching. Today, India boasts of successful launch vehicles like PSLV for Low Earth Orbit (LEO) and Medium Earth Orbit (MEO) satellites and very soon GSLV-III for High Earth Orbit (HEO) or Geo Synchronous Orbit (GSO) satellites. The success and reliability coupled with lowest launch costs in the world market has put India in a leading position for commercial satellites launches. Be it the US, the most advanced space faring nation or the third tier space operating nations like Israel, South Korea and others, there is a huge market for satellite launch. As a result, India, with its successful launch capability and a proven track record of PSLV rockets, has started reaping the benefits of space commerce in the satellite launch segment.

Moving ahead with élan, India launched a road map for space activities and finalised a blue print in the form of Vision 2020 for space. This chapter outlines many ambitious space projects which

---

3   Ashley J Tellis and Sean Mirski in "Crux of Asia – China, India and the Emerging Global Order" p 174

are part of the Vision 2020 for ISRO, which when achieved, will put India into the elite group of space faring nations. This will also enable India to utilise space in the best possible way, for not only trade and commerce, but also for strategic and tactical planning.

Indian Space Objectives

Western analysts seem to believe that India's resurgent space programme is a reaction to China's growing success in the field. But Indian officials deny this, saying its programme is entirely based on India's own needs. Also, China's programme is backed and funded by its armed forces and is therefore more defence oriented than India's. [4]

-Mr *Radhakrishna Rao, a Defence and Space Analyst*

For quite some time now, the Western space analysts have been projecting a view that India is competing with China in a space race. India's Chandrayaan-1 mission to moon in October 2008 and China's first lunar probe Change-1 in Oct 2007, both have been described as exercises to boost the national prestige of the two most populous Asian countries. Both missions had almost similar scientific objectives. While China's Change-1 terminated its 16-month mission after impacting on the lunar surface on 01 Mar 2008, Chandrayaan-1 went on to study the lunar features in addition to exploration for the presence of water and Helium-3 on the moon. Both countries, infact have an eye on the resources available on moon. While India has hinted at its eventual aim of mining for lunar resources, China is thinking of setting up a base on the moon. Interestingly, both India and China have announced plans to send a landing mission to the moon in this decade. While China is edging close to finalising its plan for a manned landing mission, the Indian Space Research Organisation has made it clear that a manned moon landing project would be taken up only if it is 'totally justified'.[5] India has further stated that they are in no way in any 'space race' with China. As per senior space scientists, Indian space programme is governed by the Indian requirements in space and its national

---

4   Dr Radhakrishna Rao, a freelancer on Defence matters

5   Article : " Is there a Sino-Indian Space Race" in "Aviation & Aerospace" May 2009 Issue

interest.

Contrary to western claims about an impending 'space race' between the two Asian giants, a look at the two space programmes gives a different perspective. China is in a different league, much ahead of India. The Chinese space programme is inching closer to the US space programme. Few space analysts predicted that the pace with which China's space programme is progressing; they will very soon overtake the US space programme and become the leading space faring nation of the world. The writing is on the wall, as the Chinese have already overtaken the US in launching higher number of satellites in the last two years, being second only to the Russians. Based on interactions with Chinese space officials, Space Daily a web portal has reported that "China has set a target of completing a space mission of '100 rockets, 100 satellites' between 2011 and 2015".[6]

China has so far accomplished successful manned space missions and created history of sorts in such a short period of its space programme. The Chinese space achievements are indeed laudable. As regards its manned space missions, ever since the launch of its first astronaut Yang Liwei in space in 2003, the Chinese have kept the pace of manned missions and achievements. In September 2008, it carried out its third such mission on the Shenzhou-7 spacecraft, which was conspicuous for the widely publicised space-walk by its astronaut Zhai Zhigang. It launched Shenzhou-8 in Nov 2011 for docking with a free-flying research module called Tiangang-1. Eventually, on 24 Jun 2012, three Chinese astronauts successfully completed a manual docking between the Shenzhou-9 spacecraft and the orbiting Tiangong-1 lab module, the first such attempt in China's history of space exploration. The successful launch of Shenzhou-10 manned mission on 11 Jun 2013 with a three member crew re-validated its capabilities of manned flights and docking capabilities in space. As envisaged now, Tiangang-1 would be a major step towards putting in place an autonomous orbiting complex. It means China has mastered the docking technologies and it is fully capable

---

6 Article published in Space Daily, a web based news portal on space, titled "China to launch 100 satellites during 2011-15" on 14 Mar 2012 accessed through http://www. spacedaily.com/reports/China_to_launch_100_satellites_during_2011_15_999.html

of transporting humans and cargo to a space station/module. This will enable Chinese to have their own fully functional space station by 2020. While China hopes to launch a space station by 2020, India at this point appears to have no specific plan for putting in place an orbital complex and even the manned mission is nowhere in sight. Thus a space race in Asia does not stand to reason.

Western space analysts believe India's plan for a manned flight in 2015 is in reaction to China' progress in this area. As things stand now, ISRO plans to send a two-person space capsule complete with life support systems, emergency mission abort and crew rescue provisions on board a version of the three stage GSLV-MKII vehicle. As part of this programme, ISRO has tied up with Indian Air Force's Institute of Aviation Medicine (IAM), to work towards setting up a well-equipped crew training facility in Bangalore. While ISRO has already come out with a design of the capsule for the proposed manned flight, how it progresses or obtains the necessary technology is unclear. However, it is speculated that India will take Russian assistance for its manned flight.[7]

The way Indian and Chinese space programmes have evolved in such a short time as compared to the space programmes of the US or Russia (erstwhile USSR), is remarkable. Apparently, the key point of their space programme is patience, self-reliance and practical application of space technology. Both countries went on to achieve goals that improve the lives of ordinary people. ISRO took up commercial deals through Antrix for space commerce with Europeans, Africans or Americans, the thrust being generation of revenue to sustain it space programmes. Another significant aspect of ISRO is that it has managed to retain its engineers, who are in charge of its space programme and has achieved the distinction of being one of the few organisations in the country with highest retention of their work force. The outcome was very encouraging and the achievements are laudable in terms of robust homegrown satellites and launch vehicles. Similarly, China's space program progressed in a short time frame, from the launch of its first satellite in 1970 till today, where it has numerous achievements in space

---

7   Future Space Programmes of ISRO as per their website http://www.isro.org/scripts/futureprogramme.aspx#Human

technology, comparable to well established space powers. So much so that, in few space technologies China has established itself as a world leader. These achievements in space technologies spells out clearly that India and China have the means, motive, and opportunity to become the technical leaders in the years to come.[8]

What really shocked India was China's shooting down of one of its own weather satellites in January 2007, once again placing it alongside Russia and the United States as the only countries capable of such a feat. By comparison, India does not yet have a single dedicated military satellite, relying instead on the dual-use telecommunications satellites for surveillance and reconnaissance. Indian fears of being left behind grew even more acute in February 2008, when the US also shocked by China's test, shot down a satellite that it said posed a threat as it fell to Earth. This covert act by the US is perceived by many as revalidation of its ASAT capability of the Cold War era.

As perceived by few space analysts, testing ASAT technology would undermine the Indian Government's long-standing opposition to the "weaponisation of space", which may spark a global arms race. However from the Indian perspective, this race has already begun. Lieutenant-General HS Lidder, Chief of Integrated Defence Staff cited his apprehensions while addressing a conference in 2008 in Delhi, "*We may get sucked into the inevitable military race of space-based applications in warfare and protection of space assets. In a life-and-death situation, only space resources would provide advantage to any military force in the future.*" [9]

History and Organisational Structure of Indian Space Programme

The Indian space programme and its policies on space activities are under the overall responsibility of the Space Commission of Government of India. The Space Commission formulates guidelines and policies to promote the development and application of space

---

8 "India takes on old rival China in new Asian space race" by Jeremy Page in Times Online, June 20, 2008

9 David R. Sands in "The Washington Times" on 25 June 2008 titled "China, India hasten arms race in space"

science and technology. Towards that, the Space Commission is supported by other national level committees, such as INSAT Coordination Committee (ICC), the Planning Committee on Natural Resources Management System (PCNNRMS) and the Advisory Committee on Space Sciences (ADCOS).

The Department of Space (DOS) under the Space Commission was created in 1972, as the implementing arm of the Space Commission's policies. The Indian Space Research Organisation (ISRO), under the guidance of DOS, is the primary space agency of India which is responsible for implementation of the national space programme. ISRO coordinates and executes all the Indian space programmes, such as the development of satellite communication, earth observation, launch vehicles, space science, space-industry development and support to disaster management. ISRO is also active in international cooperation and other tasks related to the implementation and coordination of the national space programme.

Another key player in India's space programme is Antrix Corporation Limited. This company is the commercial arm of DOS and is responsible for marketing and international promotion and exploitation of products and services related to the Indian space programme. It markets subsystems and components for satellites, undertakes contracts for satellites to user specifications, provides launch services and tracking facilities and other related services and activities.

In close collaboration with ISRO, several other specialised establishments operate under the DOS. These establishments, which are located in various parts the country, have the responsibility of different fields of the Indian space activities. The main space centres in India are:

- **Vikram Sarabhai Space Centre (VSSC), Thiruvananthapuram (Kerala)**. It specialises in the development of satellite launch vehicles and sounding rockets.

- **ISRO Satellite Centre (ISAC), Bangalore (Karnataka)**. It is the main centre for satellite development, covering structures, thermal systems, spacecraft mechanisms, power systems and satellite integration.

- **Satish Dhawan Space Centre (SDSC), Sriharikota (SHAR) (Andhra Pradesh).** This is India's prime launching pad facility, providing the launch infrastructure as well as solid propellant processing and their testing. A second launch pad has been recently built at SDSC SHAR.

- **Liquid Propulsion Systems Centre (LPSC).** Located at Bangalore, Thiruvananthpuram and Nagercoil, the LPSC handles testing and implementation of liquid propulsion control packages and helps develop engines for launch vehicles and satellites. The testing is largely conducted at Mahendragiri. The LPSC also constructs precision transducer.

- **Space Applications Centre (SAC), Ahmedabad (Gujarat).** This centre specialises in the development of payloads for communication, meteorological and remote sensing satellites. It also conducts space applications research and development. It additionally operates the Delhi Earth Station.

- **ISRO Telemetry, Tracking and Command Network (ISTRAC).** It is headquartered at Bangalore and is responsible to provide mission support to low-Earth orbit satellites and to launch vehicle missions. It also caters for software development, ground operations, Tracking Telemetry and Command (TTC). ISTRAC has Tracking stations throughout the country and also abroad at Port Louis (Mauritius), Bearslake (Russia), Biak (Indonesia) and Brunei.

- **Master Control Facility (MCF), Hassan (Karnataka).** This is India's monitoring and control centre for the geo-stationary satellites. It handles Geo-stationary satellite orbit raising, payload testing and in-orbit operations are performed at this facility. The MCF has earth stations and Satellite Control Centre (SCC) for controlling satellites. A second

MCF-like facility named 'MCF-B' is being constructed at Bhopal.

- **ISRO Inertial Systems Unit (IISU).** This unit carries out research and development in inertial sensors and systems and allied satellite elements.

- **National Remote Sensing Agency (NRSA), Hyderabad (Andhra Pradesh).** This is an autonomous institution under DOS and is responsible for acquisition, processing and distribution of data from remote sensing satellites. It applies remote sensing to manage natural resources and study aerial surveying. With centres at Balanagar and Shadnagar, it also has training facilities at Dehradun in form of the Indian Institute of Remote Sensing.

- **Indian Deep Space Network, Bangalore (Karnataka).** This network receives, processes, archives and distributes the spacecraft health data and payload data in real time. It can track and monitor satellites up to very large distances, even beyond the Moon.

The Indian space programme is basically a civilian programme under the Government of India. Hence its activities are primarily devoted to societal good and welfare of masses. There is no military involvement of any kind in the Indian space programme. This can also be seen from the organizational structure shown below:

ORGANISATION STRUCTURE OF INDIAN SPACE PROGRAMME

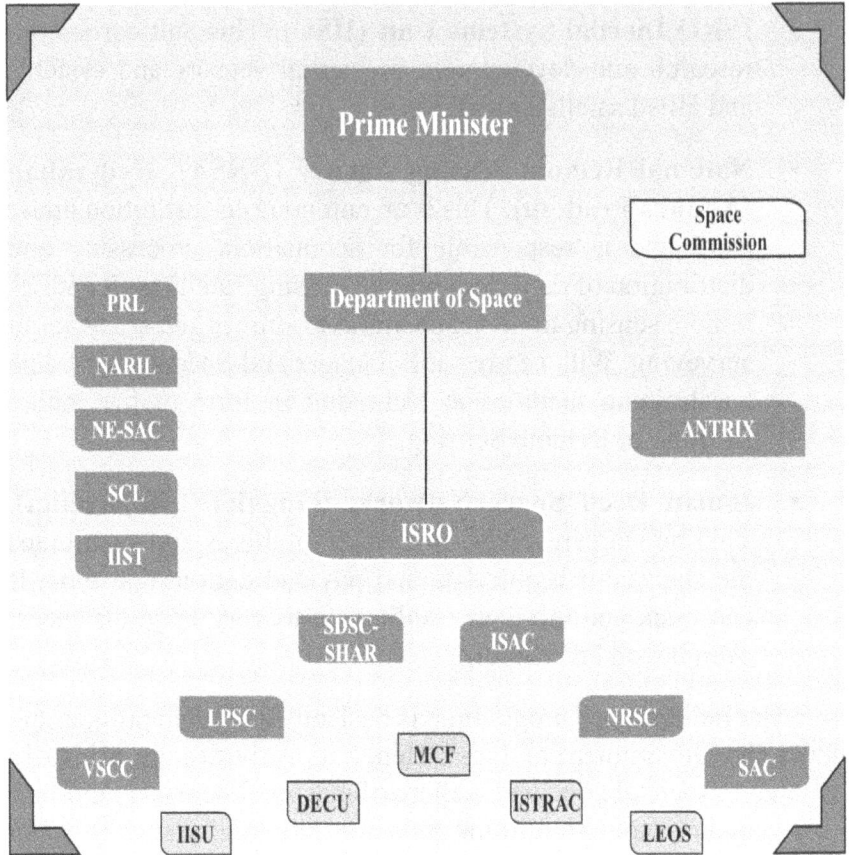

Figure – II (courtesy ISRO) [10]

The abbreviations used in the organization chart have been expanded below:

VSSC   -   Vikram Sarabhai Space Centre, Thiruvananthapuram.

ISAC   -   ISRO Satellite Centre, Bangalore.

SDSC   -   Satish Dhawan Space Centre SHAR, Shriharikota.

---

10   Accessed   through   http://isrohq.vssc.gov.in/isr0dem0v2/index.php/about-isro organisation.

LPSC	-	Liquid Propulsion Systems Centre, Valiamala, Mahendragiri and Bangalore.
SAC	-	Space Application Centre, Ahmedabad.
NRSC	-	National Remote Sensing Centre, Hyderabad.
ISTRAC	-	ISRO Telemetry, Tracking and Command Network, Bangalore.
MCF	-	Master Control Facility, Hassan and Bhopal.
IISU	-	ISRO Inertial System Unit, Thiruvananthapuram.
LEOS	-	Laboratory for Electro-Optic Systems, Bangalore.
DECU	-	Development and Educational Communication Unit, Ahmedabad.
RRSC	-	Regional Remote Sensing Centres, East, West, North, South and Central.
PRL	-	Physical Research Laboratory, Ahmedabad.
NARL	-	National Atmospheric Research Laboratory, Gadanki.
NE-SAC	-	North-Eastern Space Applications Centre, Shillong.
SCL	-	Semi Conductor Laboratory, Chandigarh.
Antrix	-	Antrix Corporation Limited, Bangalore.

Indian Space Programme : Making of a Space Power

On 07 Mar 2007, the then Indian President A.P.J. Kalam while addressing the Indian Air Force President's Fleet Review at Chandigarh Air Base envisioned bold goals for India's aerospace future. He stated that:

> *"I visualize the Indian Air Force of 2025 to be based on our scientific and technological competence in the development of communications satellites, high-precision resource mapping*

satellites, missile systems, unmanned supersonic aerial vehicles and electronics and communication systems." [11]

As per his predictions, by 2025 IAF will be a model force for the rest of the world, which will be able to "succeed in the electronically controlled warfare in the midst of space encounters, deep-sea encounters, and ballistic missile encounters." [12]

Though, national economic development has been the primary goal of India's space program from the very beginning, the military requirements were appreciated with the passage of time and they were subsequently dovetailed without compromising the basic tenets. Initial programming decisions were focused on building indigenous capabilities for building and launching satellites, thereby making the national space programme self reliant. The father of Indian Space Programme" Dr Vikram Sarabhai, steered India's space programme in this direction from the very beginning.

The defence sector of Indian space technology is handled primarily by the Defence Research and Development Organization (DRDO), which functions under the Ministry of Defence. India's missile technology has been developed and operationalised for production by DRDO. However ISRO's institutional preferences and the limits of its civilian mandate have ensured that space is not militarized and Indian Space Programme continues un-hindered towards civilian purposes, which gives ISRO to identify itself as a civilian organization. Though it consistently denies any interaction with DRDO, there are indications that ISRO does cooperate on military objectives, which are essential in the national interest. ISRO's launch of increasingly capable imaging satellites with military implications is one such indication, and some sources have even hinted the satellites were built by DRDO. [13]

---

11 Former President Kalam "IAF will be Model for World by 2025:," *The Times of India*, March 7, 2007.

12 Ibid

13 "India Building A Military Satellite Reconnaissance System," *Defence Industry Daily*, August 10, 2005

Cooperation between ISRO and military/defence elements is generally assumed to   exist. [14] For instance, the CartoSat series of satellites are purely ISRO projects for civil use, but they also have military applications.[15] The systems that are for civilian use may serve as a test bed for future military projects. Many leading space powers do follow this approach like the French SPOT system which evolved into the Helios military satellite. Even the organizational structure of ISRO (at Figure – II) has been formulated to ensure civilian control over the country's space programme. ISRO is controlled by the Space Commission, which reports directly to the Indian Prime Minister and Minister of State. It is the Space Commission which is responsible for high level policy formulation and budgetary approval for ISRO and all programs that are constituent elements of the Department of Space. [16] The primary motivation of Indian space programme is the in-house development and advancement of space technologies and capabilities. It seems, the ISRO leadership is now adopting a practical approach towards the application of space technology, wherein the onus of the utility of the satellite system is not there with them. Once a satellite is produced by them, it becomes a national asset which could be used for civilian application for national development, commercial exploitation for space commerce, or military application towards the securing the national interest, as the need be. [17]

Indian Vulnerabilities in Space

Brahma Chellany, a professor of strategic studies at the Center for Policy Research in New Delhi said, *"Whatever may have been China's motivation for their anti-satellite (ASAT) weapon test in Jan 11, 2007, it is bound to have lasting global impact like no other military event in recent years"* [18]. Though China's ASAT test was

---

14 Dinshaw Mistry, "India's Emerging Space Program," 161-7.

15 Cartosat-1 is reported to have "two cameras able to point at an object from two different angles," and Cartosat-2 operates at a one-meter resolution, and a 120GB storage for captured images.

16 Jerrold. F. Elkin and Brian Fredericks, "Military Implications of India's Space Program," *Air University Review*, May-June 1983.

17 Dinshaw Mistry, "India's Emerging Space Program," 162.

18 "India's vulnerability bared" by Brahma Chellaney in The Japan Times (on line edition) Feb 09, 2007 accessed through http://www.japantimes.co.jp/text/eo20070209bc.html

basically targeted towards seeking an asymmetric advantage over a much stronger and technically sophisticated USA, it is of much greater concern to its immediate neighbours in Asia like Japan and India, as both operate significant number of satellites and have contentious relations with China. The only counter to ASAT weapons is a capability to field an ASAT weapon for retaliation. The US and Russia have the capability to cripple China's communications and expose its ground assets if their space assets were struck. Japan has taken a very serious note of the Chinese ASAT test, and has the solace of the US security umbrella to defend its space assets. However, India on the contrary has to fend for itself against China. In order to deter and counter the Chinese, India has thus managed to test Agni-V, the missile with a reach to the Chinese heartland for a counter-offensive. But it is for sure, as of now it does not have a counter ASAT capability to deter China from destruction of its space assets, which is a matter of concern in the space security architecture. In the 21st century, fighting a war with one's key space assets disabled will be suicidal. Such assets are critical not just for communications but also for imagery, navigation, interception, missile guidance and delivery of precision munitions.

It is not lack of resources but a reluctance to get its priorities right that has left India far short of meeting its minimal-deterrence needs against China. Rather than preparing to fight a war, India ought to give greater priority to prevent a war. However to prevent war, a potent and credible deterrence is essential and inescapable. India thus needs to develop asymmetric capabilities in space to build up a credible and potent deterrence against China. This would definitely hedge peace in the Indo-Chinese relations, as a deterred China will definitely be more cautious in dealing with a resurgent India with asymmetric capabilities that can cripple the Chinese space capability and thus affect the very war waging potential of China.

Military Motivations of the Indian Space Programme

Two major military priorities emerge out of India's current space capabilities. One is improvement of satellite reconnaissance capabilities, and the second is an integrated aerospace defence command. These efforts were further prioritised after China's ASAT test. However, India did take cognizance of the Gulf War I and the use of space based services towards war fighting by the American forces. As a result, both aspects have been on the agenda of the Indian strategic and scientific community since the late 1990s.[19] But the impetus for military applications of space was received after the 1999 Kargil conflict. India's growing regional role in the world affairs and its new found economic status impacted positively towards Indian aspirations to infuse its available space technology into war fighting machinery. In this regard, the powerful impact of space assets on military powers, particularly in the US and China, and the ambitions of other Asian nations are noted by Indian defence officials. Wing Commander K. K. Nair, an IAF officer who specializes in the role of space in India's defence establishment, makes a case for India to recognize space-based assets as a modern necessity for ground, sea and air warfare, and recommends that India should plan accordingly. He attributes much of the success in Kosovo to the use of space-based assets by the US led coalitions. He states:

> *"While Kargil was characterized by lack of information in all aspects ranging from intelligence on enemy locations to targeting information, weather inputs, etc., Kosovo was characterized by a surfeit of space-based military information for the coalition forces which paved the way for nuanced application of military power and consequently decisive success in battle. If India had harnessed its extant space capabilities for national security in 1999, enemy incursions could have been detected and assessed with more accuracy; air power could have been applied with greater precision; and the loss of life would have been decreased"[20]*

It thus becomes necessary for India to go full steam towards military applications of its existing space-assets and technology and shall also plan for ambitious space technologies as a part of its modernization plan for the armed forces. Now that China has already demonstrated its ASAT capability, thereby posing a challenge to the Indian space assets, India has to not only consider the full spectrum militarization of space, but needs to develop counter offensive

---

19 Dinshaw." Mistry, "The Geo-Strategic Implications," 1039.

20 Nair, K.K, *Space the Frontiers of Modern Defence* (New Delhi: Knowledge World, 2006), p 17-18.

and defensive space capabilities to deter China against any mis-adventure and interference with the Indian space system.

However, the crucial decision to follow China and the United States in demonstrating ASAT capabilities will be contrary to the espoused view of ensuring the peaceful uses of outer space. Even before the Chinese ASAT test, the Chief of the Indian Air Force, S. Krishnaswamy had once stated, *"Any country on the fringes of space technology like India has to work towards such a command as advanced countries are already moving towards laser weapons platforms in space and killer satellites."* [21] The psychological elements of prestige in missile and nuclear technology have long been noted in India. While the civilian nature of India's space program has traditionally been its national pride, the ability to demonstrate technological advances and expertise may also be a consideration for Indian leadership to seek parity with China's advances in space technologies. Since the US and China have already demonstrated their ASAT capabilities, it is a necessity perceived by the Indian defence planners to re-work the strategic calculus post 2007 ASAT test and integrate the available technologies to develop and demonstrate its ASAT capability.

Militarisation of Space by India : The Feasibilities and Possibilities

Besides economic utility, the Indian satellites have transformed the war fighting capabilities and thus compelled the rewriting of military doctrines. Man's quest for "The Higher Ground" for tactical advantage has today reached the realms of space. If World War II triggered the momentum of satellite launch, the Cold War saw the phenomenal rise in launch of satellites and experimenting with various military capabilities of satellites, which had direct bearing on war fighting capability of Cold War adversaries. Though the economic spinoffs have accelerated the growth of satellites, yet the military utilization of space assets has been increasing with every passing year. From limited utilization of satellites for ISR role, today the war waging capability of any nation is heavily dependent

---

21 Jeffrey Lewis, "What if Space were Weaponized? Possible Consequences for Crisis Scenarios" (Center for Defence Information, Washington, D.C., July 2004) 29.

on space assets, which has transformed the war doctrines. Today C4ISR and PNT operations have seen a major transformation from manual and land based operations to more robust, prompt and reliable satellite based operations. This has made space "The Ultimate High Ground" for war fighting in the modern era. Today the satellites are being used for secured encrypted communication, navigation of aircrafts, ships, combat vehicles and even the foot soldiers, weather forecast for tactical operations, 2D and 3D maps for military planning, GPS assisted precision targeting and bombing by fighter and bomber aircrafts, positioning and timing for co-ordination of global operations, providing distress locating services, so on and so forth. The level of sophistication and dependency on satellite based services for military operations in the 21st century can be appreciated from the successful execution of "***Operation Geronimo***" by the US Navy Seals to eliminate Osama Bin Laden in Abbotabad in Pakistan. This operation was controlled from USA, and watched live at the Operations Room by the President and his select team. The entire networking, communication and transmission of data in 'real time,' all could happen so meticulously only with help of space based assets.

Thus space based assets have found great utility in the realms of military planning and operations, than they were envisaged for. Satellites can be called the RMA of 21st century. The utility of space based assets was highly visible in Gulf War I, wherein the US forces extensively utilised the satellite based services to demonstrate their "awe and strike" strategy against the Iraqi forces. The outcome of this war was swiftly decided with minimal loss of human life from American side. Even the co-lateral damage was controlled to a large extent. Rightly so, the Gulf War I has been rightly called **'The Ist Space War'**, just for the fact that never before in history were so many satellites utilized towards war fighting. The 'Op Iraqi Freedom' and thereafter the Kosovo War, Afghan War and the very recent swift action in Libya, saw increased dependence of Armed Forces on satellites. This led to extensive militarization of space by the space faring nations for conduct of war. The US and Russia were the leaders in militarisation of space during the Cold War period, but now China is catching up fast with these established space powers. The II-Tier space players like India and Japan have initiated efforts to work towards space security architecture to secure their space based assets and ensure un-interrupted utility of the space assets for securing their national interest, both economic and military.

Making of "Indian Space Security Architecture"

Space is and will remain an important arena to support future military operations. The integration of space support systems with other war fighting systems will be crucial for India in the coming years. Information domination coupled with space based assets would become an important factor that could determine the outcome of future wars. Satellites are increasingly becoming dual-use (for both economic and military purposes). Learning from the experiences of Kargil War and appreciating the exploits of space technology by the US and NATO forces in Iraq/Kosovo, the Indian scientific community and the defence establishment embarked upon a path to consolidate existing space technologies in line with the US and Russia and European nations. The Indian space programme which was dedicated all these years for social and development causes, saw it being utilized by the armed forces. India is increasingly taking steps to exploit space assets to enhance its operational capabilities in order to support network centric warfare. However the actual exploitation of space should be seen in terms of overall indigenous development and induction of space systems. The requirement of the defence forces, both in terms of communications and imagery, is presently being met through ISRO's dual use satellites. Similar arrangements can continue towards military navigation, as and when India's Regional Navigational Satellite System becomes operational.[22] However, there exists a fair amount of threat to our space assets, especially in view of the Anti-satellite (ASAT) capability developed by China, which warrants mitigation efforts by India on priority.

Keeping in view various offensive counter space technologies

---

22 Bhaskaranarayana a senior scientist of ISRO says that Antrix provided these services only on a commercial or civilian basis, and not for defence purposes. Defence services may use the data, he says, but Antrix does not offer any specific services for them. Antrix recently launched CARTOSAT-2, which offers the facility to receive data products to international users accessed theough http://battakiran.wordpress.com/category/isromilitary-missiles/.

already developed or under development in the Indian neighborhoods, the emanating threats to space systems are listed below:

- KKE ASAT attack against satellites.

- Jamming of Satellites.

- Blinding/dazzling of satellites.

- Space Mining against satellites.

- Jamming of command and control systems/links of satellite system.

- Co-orbital ASAT attacks on satellites.

- Cyber-attack on space systems.

- Physical attacks on satellite ground stations.

- Pellet cloud attacks on low-orbit satellites.

- High-altitude nuclear detonations (HAND).

- Directed Energy Weapons.

- Space debris.

In order to counter and neutralize these threats to the Indian space assets listed above, India has also started work on various space projects. The projects which are being conceptualized, developed or are being implemented towards developing offensive and defensive space capabilities will be need to be integrated and weaved together to form **"Indian Space Security Architecture."** Few of these projects have been discussed in succeeding paragraphs.

Indian ASAT

After successful testing the over 5,000 km Agni V missile, which had a range of 600 km into space during in its parabolic trajectory, the Defence Research and Development Organization (DRDO) now feels it can design and produce an anti-satellite (ASAT) weapon expeditiously. The Agni –V missile system indeed turned out to be a game changer, both as a potent ballistic missile system and as a precursor for an Indian ASAT. In an interview to Sandeep Unnithan, Senior Editor, India Today Scientific Advisor to Defence Minister and DRDO Chief Dr V K Saraswat stated,

"Today, India has all the building blocks for an anti-satellite system in place. We don't want to weaponise space but the building blocks should be in place. Because you may come to a time when you may need it. Today, I can say that all the building blocks (for an ASAT weapon) are in place. A little fine tuning may be required but we will do that electronically."[23]

The Times of India quoted Dr Saraswat, who was responding to media after the successful launch of Agni-V,

"Agni V's launch has opened a new era. Apart from adding a new dimension to our strategic defence, it has ushered in fantastic opportunities towards developing ASAT weapons and launching mini/micro satellites on demand. "An ASAT weapon would require to reach about 800km altitude, Agni V gives you the boosting capability and the 'kill vehicle', with advanced seekers, will be able to home into the target satellite,"[24]

It is thus apparent that the Indian Defence Scientists have built up the desired ASAT technology in bits and pieces, which is currently being used in AGNI missile system and other projects. The successful testing of Agni-V and Ballistic Missile Defence (BMD) has proven that the technology required for development and integration of an ASAT system is available in India. In present scheme of things the ASAT weapon would include marrying Agni V's propulsion system with the 'kill vehicle' of the under-development two-tier BMD (ballistic missile defence) system, which has already been tested and is likely to be deployed soon in Delhi, to start with. The technology of Agni –V would provide the necessary height and reach to the ASAT system and the technology of Advance Air Defence missile, with its seekers would hit the desired target in the space. Though, as

---

23 In an interview with Sandeep Unnithan, Senior Editor, India Today, accessed through http://indiatoday.intoday.in/story/agni-v-drdo-chief-dr-vijay-kumar-saraswat-interview/1/186248.html

24 Rajat Pandit in TNN Apr 21, 2012 available at http://articles.timesofindia.indiatimes.com/201-04-21/india/31378237 1 asat-anti-satellite-agni-v

of now there is no government nod for any kind of ASAT project, but there is a re-thinking of the entire issue in view of the Chinese ASAT test in 2007.

Since the country's economic development is dependent on satellite based technologies, in the existing security scenario and testing of an ASAT by China, the Indian space assets need to be secured from physical damage and destruction. Since the technology apparently is available within the country, as has been claimed by scientists from DRDO, the time has therefore come for India to initiate its effort towards demonstrating its ASAT capability, which would act as deterrence to our adversaries from initiating any attack on Indian space assets. The physical demonstration of ASAT capability is in the interest of the country, as "deterrence is potent and credible only when it is physically demonstrated".

Radar Imaging Satellite (RISAT)

India's Polar Satellite Launch Vehicle blasted off on Apr 26, 2012 with a radar surveillance satellite called RISAT-I. It is designed to obtain all-weather, day-and-night ground imagery for national security and environmental applications. The satellite marks a significant leap for India in earth observation capabilities. It is the country's first indigenous radar observation satellite after a successful series of optical and infrared imaging missions. The RISAT–1 spacecraft's C-band radar will see through clouds and take pictures of the ground at night, offering a significant leap in technology and reconnaissance data over existing optical imagers. Because they do not require sunlight to illuminate the ground, radar pulses emitted from orbiting satellites can resolve objects on Earth's surface in darkness and in all weather conditions.[25] RISAT-1 became India's first satellite with homemade radar technology. This satellite will circle earth 14 times a day with a ground track repetition every 25 days.

The analysis and inferences from these statements gives an indication that India proved its technological prowess in space technology with the launch of its indigenous RISAT – I satellite. Ii is a state-of-the-art Active Microwave Remote Sensing Satellite

---

25 Stephen Clark in Spaceflight Now on April 26, 2012 accessed through http://www. space.com/15440-india-rocket-launch-surveillance-satellite.html

carrying a Synthetic Aperture Radar (SAR) that will operate in the C-band.[26] SAR, which gives the RISAT-1 its roving eyes, provides clearer images of earth at all times and under any condition. It fires microwaves, which is reflected by the earth's surface and the SAR radar inside the satellite uses those reflections to make pictures so clear, even from 600 Km above the earth's surface. Because it doesn't need light, RISAT works both by day and night, in cloudy conditions and crucially; they see you even under trees or forests cover. That makes RISAT very useful for the military operations. This dual use capability of RISAT needs to be further improved upon for tactical military utility, wherein the requirement is to have shorter 'revisit cycles' over a desired area and better resolution (less than 50 cm).

Integrated Space Cell

In the 21st century, the Indian armed forces have to prepare for a war in yet another frontier, the outer space. Towards this, though a small, but a well thought of beginning has been initiated by the Ministry of Defence by creating an Integrated Space Cell (ISC) under HQs IDS in Feb 2008. The ISC is jointly operated by the country's armed forces, works in close co-ordination with the DRDO, civilian Department of Space and the Indian Space Research Organization (ISRO). This strategy will ensure effective utilization of the country's space-based assets for military purposes and to look into the possible threats to these assets, as well.

The Integrated Space Cell has identified and projected to the Ministry of Defence, three space-based systems necessary for India's current and future defence requirements:

- **Imagery and surveillance capabilities**. The remote sensing satellites almost always perform dual-use functions. In the past, data from the Indian remote sensing series of satellites have had several clients, both domestic and foreign. For example, the Indian Army has used the Indian Remote

---

26 Varma M Dinesh wrote in "The Hindu" on 26 Apr 2012, "PSLV-C19 puts RISAT-1 in orbit" accessed through http://www.thehindu.com/news/national/pslv19-puts-risat1-in-orbit/article3355368.ecehttp://www.thehindu.com/news/national/pslvc19-puts-risat1-in-orbit/article3355368.ece

Sensing series of satellites' data for mapping purposes. An important technological component of the Indian imagery and surveillance satellites is the synthetic aperture radar satellites, which guarantee coverage under cloud cover. Though India has been able to develop this niche technology, but its space-based early warning capabilities are limited. Further, Indian agencies are limited in their ability to integrate data from information gathered by separate terrestrial, air, land, and seaborne sensors. This gap perhaps will be bridged consequent to the operationalisation of Naval Satellite (GSAT-7).

- **Positional and Navigational Capabilities**. India's aerial delivery systems (air, land and sea) use the US GPS. However, this system produces large errors due to the large height difference between the Himalayas (height of 8,000 meters) and the Indian Ocean (depth of 4,000 meters). Hence, the GPS-aided geo-augmented navigation system (GAGAN) will be used to provide an automatic correction to GPS signals. Also, there is a need to have an alternative to the GPS system, which for strategic reasons needs to be a constellation of Indian satellites, which will be fulfilled once the Indian Regional Navigation Satellite System (IRNSS) gets into action. The first satellite of this planned constellation is scheduled to be launched in 2013. This constellation of IRNSS satellites will be able to provide India an independent regional navigation system which will cater the needs of Indian operators in both, the South Asian landmass and the Indian Ocean.

- **Reliable Communication Capabilities**. The Indian armed forces have been using the Indian National Satellite System (INSAT) series for communication. However, ISRO aims to increase the number of communication satellites significantly. But growing population with significantly expanding needs in commercial communication capabilities and additional communication requirements armed force's networks to enable communication with assets such as unmanned aerial vehicles and multi-role aircrafts, will call for additional bandwidth for reliable and secure communication. Further the induction of India's submarine-launched ballistic missiles will further increase India's communications requirements. The Department of Space, thus has to plan and ensure that these requirements are optimally met.

The Integrated Space Cell (ISC) was created under HQs IDS in Feb 2008 in order to counter the increasing threat to India's space assets. Unlike an aerospace command, which is service-specific, the ISC envisages cooperation and coordination between the three services as well as civilian agencies dealing with space. The Indian Army, Air Force and Navy will therefore work together in the Integrated Space Cell and coordinate with each service HQs for optimal utilization of the available space-based assets.

The satellites are also a vital link in India's defence and security architecture. For all its impressive achievements in building and launching satellites, India is now trying to find ways and means to defend them. If these satellites were to be damaged or knocked off, India's communication network would be broken, its security severely jeopardized and its capacity to defend itself against aggression damaged immeasurably. Thus India too needs to protect or shield its satellites against ASAT technologies which could be utilized by an adversary. Dr Lawrence Prabhakar, a noted space analyst, supporting the creation of Integrated Space Cell opined that, the government has set up the Integrated Space Cell to look into the threats and challenges that India's space assets face. According to him a long-term goal of the cell would be to integrate space and ground operations for civilian and military objectives. Pointing to the different kinds of satellites, space-based laser systems, space stations and ground-based laser stations for offensive space operations that the US, Russia and China have already developed or are developing, Dr Prabhakar intimated that India's space architecture of offensive and defensive systems are yet to be conceived, built and deployed. In the event of their satellites being knocked out by enemy action during a crisis, the US, Russia and China have the capability to launch substitute satellites into space at short notice. The US can move its satellites from one orbit level to another, higher level to escape being taken out by an enemy anti-satellite system (ASAT).

But as of now, India can plan a satellite launch only on a programmed sequence basis and not on short notice for rapid launches to replenish lost satellites. But as of now, India does not even have preliminary capability to defend its satellites. It will take another 15 to 20 years or more before India can put these systems in place.

Dedicated Naval Communication Satellite

This satellite is planned to be launched in 2013 [27] for providing secure and dedicated communication network to Indian Navy. This dedicated communication satellites will be providing real time data from multiple sites to the Indian Navy. The project entails inter-connection of weapons on board all its warships. In the first phase, 20 warships have been selected and work has already begun. [28] This is aimed at achieving awareness of the maritime domain and network-centric operations by the Indian Navy.

This satellite will combine the Naval Comsat and Naval Surveillance requirements/ missions, which will eventually cover a swath of around 1000 NM over the Indian Ocean and will greatly enhance the Indian Navy's C4ISR capabilities. It will also provide a seamless flow of information through different communication network in use with Indian Navy by using data generated from multiple sources and assets such as reconnaissance satellites, UAVs, ships and personnel carrying communication & surveillance equipments, radar networks (both ground based and air borne early warning). Once operational, this satellite will be able to link every aspect Military Operation to a single point of confluence which will be able to provide the Theatre Commander a wide, accurate and a holistic overview of the battlefield scenario, thereby assisting him in the decision making. Another feature of this network is that it will be isolated from the World Wide Web (WWW), so as to provide protection from cyber-attacks.

---

27 *The Times of India* on Apr 30, 2012 " Dedicated satellites for Navy, IAF to be launched soon: Antony"

28 Asian Defence, "Satellite to Connect Indian Navy Warships" accessed through http://theasiandefence.blogspot.in/2009/10/satellite-to-connect-indian-navy.html

Indian Regional Navigation Satellite System (IRNSS)

IRNSS is India's own regional navigation satellite system being setup by ISRO to provide accurate real-time positioning & timing services over India and the adjoining region extending up to 1500 Km around India. It offers two services, Standard Positioning Service for commercial usage and Restricted Service with encryption for dedicated military usage. The fully deployed IRNSS system will consist of 3 satellites in GEO orbit and 4 satellites in GSO orbit, approximately 36,000 km altitude above earth surface. Each satellite is configured around I1K bus and is continuously monitored & maintained by ground segment. Each satellite has two payloads viz. Navigation payload and CDMA Ranging payload along with retro-reflector. The design of the payload makes the IRNSS system inter-operable and compatible with GPS and Galileo.[29]

The IRNSS will be a seven satellite constellation, out of which at least three satellites will have to be in GEO overlooking the Indian Ocean. All these seven satellites will have continuous radio visibility with the Indian control stations. The satellite payloads will consist of atomic clocks and electronic equipment to generate the navigation signals.

GPS Aided Geo-augmented Navigation System (GAGAN)

The Ministry of Civil Aviation has decided to implement an indigenous Satellite-Based Regional GPS Augmentation System also known as Space-Based Augmentation System (SBAS) as part of the Satellite-Based Communications, Navigation and Surveillance (CNS)/Air Traffic Management (ATM) plan for civil aviation. The Indian SBAS system has been given an acronym GAGAN - GPS Aided GEO Augmented Navigation. A national plan for satellite navigation including implementation of Technology Demonstration System (TDS) over the Indian air space as a proof of concept, has been prepared jointly by Airports Authority of India (AAI) and ISRO. TDS was successfully completed during 2007 by installing eight Indian Reference Stations (INRESs) at eight Indian airports and linked to the Master Control Center (MCC) located near Bangalore.

---

29 Refer ISRO's Space Application Centre download on IRNSS accessed through http://www.sac.gov.in/SACSITE/IRNSS-1A.html

This will be a step towards introduction of modern communication, navigation, surveillance/Air Traffic Management System over the Indian airspace. It will involve establishing 15 Indian reference stations, 03 Indian navigation land uplink stations, 03 Indian Mission Control Centres and installation of all associated software and communication links. Once operationalised, it will increase safety of Air Traffic Management by using 3-D approach operation with course guidance to the runway. It will provide high positioning accuracies over the Indian airspace by providing ' 3 metre accuracy'. The positional accuracy will be simultaneously available to 80 civilian and more than 200 non-civilian airports and airfields, which will eventually be increased for 500 airfields.

The Preliminary System Acceptance Testing of the GAGAN programme has been successfully completed in Dec 2010. The first GAGAN navigation payload was flown on GSAT-8 which was launched on May 21, 2011 and the second on GSAT-10 launched on 29th September 2012. The Navigation payload on GSAT-10 would provide improved accuracy of GPS signals (of better than 7 meters) to be used by the Airports Authority of India for Civil Aviation requirements. Once operationalised, the GAGAN project, apart from air traffic management in the Indian subcontinent will also accord air surveillance of Indian Air Space and facilitate in 'Air Situational Awareness' for military operations.

Military CCI-Sat Reconnaissance Satellite.

India is planning it first Communications Centric Intelligence Satellite (CCI-Sat), which is planned for launch in 2014 in a circular 300 mile high orbit. The satellite payload is valued at $25 million for the satellite program and is being built by the Defence Electronic Research Laboratory of India. But its design and research and development testing is being handled by the Indian ISRO space research administration. It is a part of a larger development for military intelligence enhancement utilizing advanced electronic warfare systems capability as applied through satellites to operate as RECSAT and SIGINT platforms for military requirements.[30]

---

30 "India's Dedicated Military CCI-Sat Reconnaissance Satellite " accessed through http://www.globalsecurity.org/space/world/india/cci-sat-recsat.htm

Dedicated Encrypted Defence Communication Network (DCN)

A facility for secure defence communication through dedicated Communication satellites is in the offing. This will be a part Network of Networks, wherein, the individual communication networks of Army, Navy and the Air Force would be linked up to form a secure Defence Communication Network (DCN) architecture. The project is being worked out through public-private partnership, which is yet another effort to shore up private partnership in defence projects. The communications and information technology is processing the proposal for the OFC network to be laid over 60,000 km to provide connectivity for 129 Army, 162 Air Force and 33 Navy stations. [31]

All Weather Tactical ISR Capability

To facilitate the Indian Armed Forces in assessing battle field scenario and planning strategies and tactical moves, there is a need to develop dedicated satellites for specific military use, preferably at a short notice. This will call for launch of satellites at a short notice for tactical ISR for the armed forces to deal with the emerging battle situations and safeguarding national interest.

Weather Forecast for Tactical Missions

To provide real time satellite imageries for planning operations, real time weather data has become essential towards planning of any military operations. Thus the weather satellites have to be integrated with military establishments right up to the field level units, to assist them in planning the military operations. The weather forecast not only needs to be packaged, but also needs to be readily made available to the operators in the field, which can maximize the impact of the planned military operations.

Micro-satellites

Talking about micro-satellites, former Secretary of Defence William J. Perry stated, "*Space forces are fundamental to modern military operations.*" [32] However, they are the most vulnerable targets for

31 "India to upgrade defence communication network" accessed through http://www.igovernment.in/site/india-upgrade-defence-communication-network

32 "Micro-satellites: Charting a New Course to Space Security" by William S. Marshall

a space faring nation. The space assets up till now were not only voluminous but as a single unit they were equipped to perform multiple functions. However, keeping in view of their vulnerability and yet their importance, it is felt that instead of a space architecture that consists of a few large satellites that are voluminous, complex and expensive, the space faring nations have to move to a new model that is "constellation-like, consisting of many inexpensive small or micro-satellites". The space systems as of now perform five main functions towards military use i.e., early warning of missile attack, navigation, communications, signals intelligence, and reconnaissance, invariably through a small constellation of big satellites. Even if one satellite of the constellation fails, the system would be significantly compromised or may even be rendered totally un-usable. Thus there is a need for the Indian policy makers and planners to consider use of micro-satellites or a constellation of micro-satellites as a replacement to the bulky satellites which are so very vulnerable to enemy action.

A **"Multi-tiered Micro-satellite Constellation Architecture" (MMCA)** will be able to perform the same functions which a large satellite is doing today, but would be less vulnerable to enemy attack. By using the MMCA, a system would be created with no room for single point failures to any known method of attack: several orbital altitudes where physical attack is difficult; constellations of satellites that are well dispersed; significant levels of redundancy; and modular satellites that work together. Such a system is possible both technically and affordably because of the general trend of miniaturization of technology and the increasing capabilities of microsatellites. Even the replacement to the damaged micro-satellites can be launched at a short timeframe.

In this context, DRDO feels that a spin-off from Agni V test is that it can work towards launching of mini-satellites for battlefield use, in case the country's main satellites are rendered useless. The

---

in *Summer* 2006 published in BelferCenter Newsletter, Belfer Center for Science and International Affairs, Harvard Kennedy School accessed through http://belfercenter.ksg. harvard.edu/publication/19285/microsatellites.html?breadcrumb=%2Fexperts%2F805 %2Fwilliam_s_marshall%3Fback_url%3D%252Fpublication%252F19285%252Fmicr osatellites%26back_text%3DBack%2520to%2520publication.

quick launch capacity and the 'on demand' capability of DRDO to launch micro/mini satellites through AGNI-V missile in any desired orbit, will be able to cater for the redundancy, in case the Indian satellites are damaged or destroyed by an adversary.

Missile Defence and International Cooperation

Dr Saraswat indicated as early as December 2007, that India will have a fully capable ballistic missile defence system to defend the nation from ballistic missile.[33] According to him,

> *"The shield has been designed with a range of 2,000 km that should be able to counter any ballistic missile threat with a reaction time of less than four minutes. The shield also is designed with two layers so that two missiles can be fired simultaneously at a target both outside the atmosphere at a 45-50 km range, and also inside the atmosphere at a 15-25 km range. "The system is being configured with radars for long-range surveillance and tracking, command, control, and communication systems for effective control and optimization of exo- and endo-atmospheric interceptors at altitudes above 40 km and below 25 km respectively."*[34]

The Indian Ballistic Missile Defence System has been discussed at length in Chapter VII.

International Co-operations Satellites : The Strategic Gains for India

Though numerous indigenous systems have been used in the past in Indian satellites, India still opted for buying a few core systems and niche technologies from the United States, Israel and Russia for providing military capabilities for its satellites. Further, India also announced the end of its IGMDP program, a signal that future missile projects will incorporate the private sector and international cooperation. [35] The countries with which India has gone into

---

33 Steve Herman, "India Plans Missile Shielf by 2010," Voice of America News, December 13, 2007.

34 "India to build anti-ballistic missile shield by 2011," Kalinga Times, January 25, 2008

35 "India Expands Foreign Collaboration in Missile and Space Program," March 2008

commercial contracts for utilization of their space technologies are discussed below.

Israel

Israel and India have cooperated on a number of ventures, as India's launch capabilities complement Israel's developments in remote sensing satellite technology. Israel's synthetic aperture radar is one such technology India successfully adapted in its space programme. This allows a satellite to "see through" clouds, rain and other adverse conditions and high resolution pictures using radar technology. The decision to launch Israel's TecSar from India was reportedly hailed by defence officials as an important milestone symbolizing the growing military ties between the two countries. [36] The launch of TecSar, designed to monitor Iran was successfully launched from India's Satish Dhawan space center in January 2008. [37] Bilateral space civil cooperation for education and exploration purposes has also increased. India has become the largest client of Israel's defence industry, and Israel India's second largest supplier (after Russia). Brig Gen (Retd) Shmuel Brom, the director of Institute for National Security Studies, an academic institute in Tel Aviv focusing on Israeli national security, finds that the growing military relationship between Israel and India is only becoming more significant. There is some speculation that these growing ties will result in explicit technology sharing deals. Joint research and development projects are also being considered as a focus area. Agreements on joint development of surface-to-air missiles (SAM) and interceptors, ship-launched rotary unmanned aerial vehicles, and advanced radars[38] are indications that Indo-Israeli defence cooperation is moving from more traditional arms deals to long-term cooperation.[39] The best and noteworthy example is the joint development on an advanced Barak medium-range SAM led by the Israel Aerospace Industries, IAF and DRDO. The DRDO is the prime contractor of this missile defence

---

36 Yaakov Katz, "Israel to Launch New Spy Satellite Aboard Indian Rocket. IAI-developed TecSar will keep an eye on Iran in all weather conditions," *The Jerusalem Post*, July 20, 2007.

37 "India Launches Israeli Satellite" *BBC News*, January 21, 2008.

38 "Israel flaunts wares and Defence Expo," *The Times of India*, February 17, 2008 .

39 P.R. Kumaraswamy, "Indo-Israeli military ties enter next stage," ISN, March 8, 2007.

system. [40] This missile will increase the current 70 km range of Israel's Barak missile, taking it up to 150 km. [41]

This high level of cooperation is important, as it shows increasing long-term interest in joint development between Israel and India on advanced weapons technology that have potential uses as space weapons. India has already imported and modified Israeli technology in missile defence. One example is the long-range tracking radar (LRTR) which was used in the "exo-atmospheric" BMD system that intercepted an incoming Prithvi missile in November 2006. This technology originated from the Israeli Green Pine radar which was imported by India and subsequently modified by DRDO to make it capable of tracking intermediate-range ballistic missiles. However, except for the basic radar, the rest of the elements were developed in India.[42] Thus, cooperation also demonstrates that India is able to take advantage of niche technologies developed outside the country, and incorporating them into indigenous systems. Israeli companies have also become very active in modernizing Soviet/Russian military equipment, which is beneficial to India. IAI has successfully integrated 'night vision capability' on to the Russian MI-35 attack helicopters. The AWACS system has been successfully built by Israel on the Russian IL-76 heavy-lift platform procured by India, which have now been operationalised in the Indian Air Force.

Russia

The erstwhile USSR and now Russia has been the largest and traditional supplier of arms to India. However, Russia and India gradually moved from a supplier-client relationship to one where joint development and marketing of weapons systems is emphasized, while protecting intellectual property rights of Russian technology through an IPR agreement.[43] The BrahMos supersonic cruise missile is one example of a joint project. Even the Soviet-era Global

---

40 "India, Israel to Jointly Develop $2.5 billion Missile Defence System," *World Tribune*, July 20, 2007.

41 M Shamsur Rabb Khan, "Indo-Israel Defence Cooperation: A Step in the Right Direction," December 23, 2007, Institute of Peace and Conflict Studies.

42 "Israeli Imprint in Prithvi Missile Test," *The Times of India*, December 3, 2006.

43 Vladimir Radyuhin, "India, Russia to Sign Defence Secretary Agreement," *The Hindu*, April 27, 2005.

Navigation System GLONASS will be used to track the missiles' movements, and as such is planned to be made fully functional by joint efforts. This development reduces Indian dependence on the US Global Positioning System. Further, the IPR agreement between Russia and India was signed in December 2005, which paved the way for future cooperation in advanced technology. On the lines of India's cooperation with Israel, these agreements will enable India to develop advanced technology and weaponry without necessarily having indigenous capability, as well as reduce reliance on any one source of supply.

United States

The United States and India continue to explore cooperation on missile defence, and even a joint missile defence shield. Secretary of Defence Robert Gates indicated in late February 2008 that the US is seeking a steady expansion of the defence relationship with India. Since 2002, there has been talk of expanding cooperation on missile defence and other advanced technologies.[44] The important moment in the Indo-US defence co-operation was in June 2005, when India and the US signed a 10-year defence pact guiding strategic cooperation on defence collaboration in the areas of two-way defence trade, technology transfers and coproduction, expanded missile defence collaboration, and a bilateral defence procurement and production group. In cooperation on missile defence, specific collaboration focused on integrating outside technology that improves guidance, fire control, and other issues is likely to be the goal.

Impact of Chinese ASAT on India's Economic and Strategic Architecture

Today India is a major space player in the world with a total investment of almost Rs 60,000 crores in its space assets.[45] These space assets have been able to propel the Indian economy as a result of which India has been able to post a positive growth of more than 6% even in the times of worst kind of economic meltdown world over due to economic recession. Besides economic utility, the Indian

44 "Joint Statement of India-US Defence Policy Group," May 23, 2002, Washington DC.

45 Sandeep Unnithan April 28, 2012 in India Today accessed through http://indiatoday.intoday.in/story/agni-v-launch-india-takes-on-china-drdo-vijay-saraswat/1/186367.html

armed forces are also reaping the benefits of satellite based services and have been effectively able to weave there strategic planning and tactical moves around the space based assets/services. The navigation, C4ISR, tactical surveillance of desired area, encrypted secure communication, weather forecast for 2D/3D maps for military operations, GPS assisted tracking, navigating and precision bombing of targets and border management are few of the satellite assisted services which have been dovetailed into the structure of the Indian Armed Forces. Hence, loss of a satellite as a result of direct Chinese ASAT attack, will not only render the military response of Indian Armed Forces tooth-less, but will also have a crippling effect on the economic activities of the country for years to come. India should therefore analyse the pros and cons of an ASAT attack from China and plan for appropriate responses to-

- Develop and demonstrate a credible and potent ASAT to deter the adversaries from launching ASAT attack against Indian Satellites.

- Develop small mission specific satellites (Microsat /Nanosat) for supporting Armed Forces Operations, as replacement to lost satellite.

- Develop quick re-launch capability to instantaneously launch small mission specific satellite for supporting military operations.

- Harden the new satellites to withstand ASAT attack to certain extent by providing camouflaging and maneuvering capabilities.

In 2011 China surpassed the US tally, by launching 19 satellites and lagged just behind Russia, which leads the world. The feat was repeated in 2012 again, which shows the frantic pace of development of Chinese space programme and thus is not only worrying the US, but also the Indian policy makers. Consequent to the testing of ASAT in 2007, China has noticeably become more belligerent in its approach towards its South Asian neighbours, including India. It has started claiming sovereignty over disputed islands in East China Sea and South China Sea by exerting force and displaying provocative stance. So much so, it has dispatched its surveillance aircraft over Diaoyu Island, which was countered by the Japanese by sending their F-16 fighter aircraft to defend its airspace over the island which they call Senkaku. The issue has the potential to flare up in the days to come, as both countries are backed by nationalist fervor. The border transgressions by the Chinese forces in Ladakh region is common, beside renewing the claim for the state of Arunachal Pradesh as part of Chinese territory linking it to be a part of South Tibet. The 19 Km incursion in Depsang Ladakh by the Chinese troops in mid Apr 2013, almost took the nation to ransom for more than a fortnight. The claim over Scarborough Shoal near Philippine's EEZ is also causing tensions in South China Sea, apart from Spratly islands. These claims and counter-claims have raised tempers in South Asia so much so, that the "hot spot " seems to shift from West Asia to South Asia. The sea lanes in the Indian Ocean are considered among the most strategically important in the world. According to the *Journal of the Indian Ocean Region*, more than 80 % of the world's seaborne trade in oil transits through Indian Ocean choke points, with 40 % passing through the Strait of Hormuz, 35 % through the Strait of Malacca and 8 % through the Bab el-Mandab Strait. [46] Any disruption in these sea lanes would cause major impact on the world trade, which in turn will create instability in South Asia, which may further add to the economic woes faced by the world. So concerned is the US with this new Chinese approach that the Obama administration was forced to renew its strategy for Indo-Pacific and has now responded with Obama's famous "Pacific Pivot", wherein 60% of the US Naval assets are now being positioned in the Indo-Pacific region to counter the evident Chinese hegemony.

India too needs to respond to belligerent Chinese moves on its borders and also in areas of national interest (including offshore) to secure its national interest and strategic balance. The Chinese PLA might have certain advantages over Indian Armed Forces in the land warfare in the Ladakh and Arunachal sectors, but Indian Navy definitely has much greater advantages in the Indian Ocean as compared to PLA Navy. No wonder the Chinese are aware of this fact and thus they have resorted to strategic encirclement of India

---

46 Sergei Desilva-Ranasinghe in "Why The Indian Ocean Matters" article published in The Diplomat on 02 Mar 2011

by creating naval bases in IOR countries, which is often referred as 'String of Pearls' strategy. This has prompted India to adopt a 'Look East Policy' not only for trade and commerce, but also for strategic engagement with the South Asian neighbours to counter its strategic encirclement by China. Keeping in view the current geo-strategic scenario and the prevalent geo-politics in Indo-Pacific region, India should respond by -

- Creating Naval facilities in South Asian countries like Vietnam, Taiwan, Australia, Malaysia, Brunei etc.

- Increase the surveillance capabilities in Indian Ocean.

- Modernise Navy for true "blue water" capabilities, with special focus on China Sea.

- Effective patrolling of Indian Ocean in order to monitor and if need be choke transition of Chinese ships (both commercial and naval).

- Development of a major naval base in Andaman Sea for swift response to a crisis in the Indo-Pacific region.

- Constructive engagement of SAARC countries to have friendly neighborhood, so as to counter strategic encirclement by China.

The Way Ahead

Having discussed the dangers of ASAT and its crippling effect in the realms of Indian comprehensive security, it is felt that the Indian strategist and the defence scientist need to come together to develop suitable technologies and strategic/tactical responses to this new threats to the National Security emanating in the region. New doctrines have to be developed by the Armed Forces to cater for the new threat and suitable futuristic requirements have to be worked out in consultation with the Defence Scientists, who shall expeditiously design, develop and demonstrate the capabilities both

- Offensive – to deter the adversaries.

- Defensive – to secure our own space assets.

Taking a cue from **"Non-alignment 2.0"** [47], a foreign and strategic policy guideline formulated for India in the twenty first century, it is incumbent on the Armed Forces to review and re-asses their military capabilities and future requirements and project the same through Ministry of Defence to Space Commission for fast track "in-house" technological development of space capabilities. Infact, the existing composition of the "Space Commission " needs to be restructured to make way for representation for Armed Forces, who will be major stake holders in defending the national space assets from any enemy attack. It needs to be appreciated that in the US, Russia and China, the handling and launching of satellite is primarily with their Armed Forces. Even if we agree that the handling of space activities in India should be with the civilian control, Armed Forces need to be on board to discuss and decide about their requirements rather than being a mute user with no say at all.

In view of the prevailing geo-political situation and the emanating threats to the Indian security, few suggestions are being made for the strategic utilisation of satellites, re-alignment of space activities and re-structuring of organizations associated with National Security.

- **Representation of Armed Forces in the Space Commission.** Unlike China, US and Russia, India's space programmes are controlled by civilians. Since the onus to protect the Indian satellite system would be with Indian Armed Forces, there is a strong need to create a **'Indian Space Security Architecture'** keeping in view the requirements of space weapons and niche space technologies. Towards this, the Armed Forces need to be taken on board, as a member of the Space Commission, for assessing its military requirements in space and projection of the same for planning, development

47 Non Alignment 2.0 is the product of collective deliberation, debate and report writing involving a diverse and independent group of analysts and policy makers, namely: Sunil Khilnani, Rajiv Kumar, Pratap Bhanu Mehta, Lt. Gen. (Retd.) Prakash Menon, Nandan Nilekani, Srinath Raghavan, Shyam Saran, Siddharth Varadarajan. The group was convened in November 2010 and met at regular intervals for over a year, until January 2012. Also present at some of the meetings were the National Security Advisor, Shivshankar Menon, and the Deputy National Security Advisors, Alok Prasad and Latha Reddy.

and validation of new technologies and integration of the same in the armed forces for effective deployment.

- **New Space Doctrine for Ministry of Defence.** The current military doctrines have been written based on current technological advancements. However, since India is now making rapid progress in space technology and is likely to utilize space for military use, there is an urgent need for all the three arms of the Indian Defence Forces to review and formulate a space based doctrine for the 21st century. This would require a paradigm shift in strategic thinking and a huge capital investment.

- **National Space Command.** This is an essential requirement for optimum utilization of space based technologies by Air Force, Army and Navy, with suitable doctrine for Offensive and Defensive Space Operations. As of now a very modest beginning has been made by creating an Integrated Space Cell under IDS, so as to co-ordinate the space requirements of armed forces with ISRO and DRDO. This has to now mature to the level of a National Space Command, to be manned by specially selected personnel from all three arms of defence services.

- **Creation of Space Force for conduct of space operations.** The space operations will be highly specialised in nature. Hence, the Indian armed forces need to create a 'Space Force', which should be adequately trained in highly technical space technologies, thereby enabling them to handle the counter space defensive/offensive operations. Towards this, a select groups of personnel have to be sent to DRDO and ISRO to appreciate the technological nuances of space based assets and for hands on training from operators perspective.

- **Training of Space Force personnel.** Indian armed forces need to take of capacity building and specialized in-house training of its 'space force', so as to enable them to handle the evolving space capabilities towards offensive and defensive space operations. The in-house training infrastructure towards training needs in respect to space based weapons and technologies have to be worked out in consultation with DRDO and ISRO and suitable training SOPs have to be formulated.

- **Strong time bound Space R&D.** The R&D departments of space establishments must start delivering projects on time. A new beginning and a focused approach for defence oriented space technologies and capabilities is the need of the day. Thus, India needs to rejuvenate and review its R&D for space weapons and space technologies. Such projects need to be planned on realistic time frames. If need be, members from Armed Forces be co-opted towards time bound management of Space Projects which have military implications and utilities.

- **Development of an ASAT capability, as deterrence and its demonstration**. Only a visible deterrence creates a desired strategic impact. A potent and a visible deterrence of an Indian ASAT capability is the need of the hour in the current geo-political scenario. The scientific community which has declared indigenous capability for building of an Indian ASAT, should not only assemble one, but also needs to demonstrate these capabilities and validate technologies. The sooner this is done, the better for the country, or else India may find itself faced with an ASAT ban regime.

- **Use of Foreign Satellites.** Using the satellites of a friendly country, especially for communication and imagery data will be advantageous. With such space co-operation, any adversary will be reluctant to target foreign satellites due to obvious political implications. Data can be thus obtained either through mutual cooperation and sharing the resource or may be purchased.

- **Electronic Warfare (EW) and Cyber Attack.** Electronic warfare coupled with cybernetic attack will make fighting wars in space both economical and effective in comparison to employing other forms of weapons against space system. Thus, India should concentrate its effort to expand the scope and capability of existing EW units to include the cyber warfare capabilities and other EW measures against the satellites by employing static, mobile and on the move satellite terminals, which will provide safety against both detection and conventional weapon attack. The pool of civilian expertise available in the field of cyber domain should be considered by the Government and mechanisms should be worked out to utilize them towards development of offensive and defensive cyber capabilities in the space domain.

- **Quick Launch and Mobile Launch Capability**. In order to replace critical space infrastructure if threatened or disabled as a result of enemy strike, quick launch capabilities through mobile platforms need to be developed for low orbit satellites. This should be planned only after carefully weighing the necessity with reference to dependability of forces on space assets, finances and other better and cheaper alternative options, such as availability of launch facilities with friendly and dependable countries.

- **Space Security Treaty for Non-offensive use of Space**. The objective of space security needs to be promoted in terms of a non-weaponised architecture, with a code of conduct regulating the pace of space activities so as to enhance the security of space assets and their non-offensive employment. This would help India in safeguarding its space assets through an International regime which would be under the patronage of the UN.

CHAPTER – VII

SPACE WEAPONS: THE INDIAN PERSPECTIVE OF ASAT AND BMD

Space weapons are weapons used in space warfare. They include all such weapons that have the capability of attacking space systems in orbit (i.e. anti-satellite weapons), attack targets on the earth from space or disable missiles travelling through space. In the course of the militarisation of space, such weapons were developed mainly by the contesting superpowers during the Cold War, and some of them are under development even today. The broad categories under which the entire spectrum of space weapons can be categorized are as follows:

- Earth-based weapons that move through space to get to targets on the ground

  - Medium to long range ballistic missiles

- Earth-based weapons that attack targets in space

  - Direct ascent ASATs

  - Lasers

  - Directed energy weapons

- Space-based weapons that attack targets in space, on the ground, or in the air

  - Co-orbital ASATs

  - Hypervelocity rods

  - Space-based lasers

PART - I

INDIAN ASAT: SECURING THE INDIAN SPACE

India's space program has very strong civil roots. It began as a means to assist India in its development and has mainly focused on improving the daily lives of its citizens. More recently, India has made a dramatic shift in the tone of its space efforts. Lately, the country has adopted a more militarized attitude, as is visible by the increased efforts by India to create an indigenous ballistic missile defence program and a thought process to shape up an ASAT programme. For example, a hit-to-kill missile defence requires developing a ballistic missile that can reach high enough to intercept a target either high within the earth's atmosphere or outside of it. Both missile defence and ASATs require the capability to quickly and accurately detect and track objects in space. But the difference lies in their type of target: a missile defence is aimed at objects moving on a parabolic flight path, whereas ASATs are intended for objects in orbit. That is a big difference, as the two targets follow different types of trajectories and travel at different speeds, and each has its own technical challenges. However, the capacity needed for either program is similar enough that a missile defence program could very reasonably be used as a technology demonstrator program for an ASAT capability. China proved this in Jan 2007 by shooting down its defunct weather satellite FY-IC through an ICBM system. The United States also revalidated its ASAT capability in February 2008, when it used a modified interceptor of its ship-based Aegis missile defence program to destroy an unusable satellite USA 193, close to atmospheric re-entry.[1]

Chinese ASAT: A Threat To Indian Space System

Amidst hundreds of space activities going on relentlessly in space, one specific incident on 11 Jan 2007, which took the world by surprise, was the shooting down of an ageing defunct weather satellite by

---

1 Victoria Samson at "Space, Science, and Security: The Role of Regional Expert Discussions," held in New Delhi, India, from January 19–21, 2011 available at http://www.thespacereview.com/article/1838/1

China. This was first successful direct ascent anti-satellite (ASAT) weapons test by China, which involved launching a ballistic missile as a Kinetic Kill Energy (KKE) vehicle to destroy one of its own defunct Fengyun-1C (FY-IC), a weather satellite at about 530 miles up in low earth orbit (LEO) in space. This unannounced testing of an ASAT was of specific relevance to the Indian space community and to the defence establishment. The Indian satellites, which have become the mainstay for its economic resurgence, were never so vulnerable to the Chinese attack.

This one incident was a game changer in the strategic planning of India. It compelled adoption of a new philosophy in Indian thinking about "management and security of space assets". The incident compelled the Indian space scientist and strategists to prepare for

- The security of defenceless satellites which are so vulnerable in their orbits.

- Ways and means to safeguard the satellites in orbit.

- Develop an Indian ASAT, as a potent and credible deterrence in response to the Chinese ASAT.

The palpable fear of competition and confrontation with China compels the Indian policy makers to rethink about their space programme meant only for peaceful and social causes for the civilian populace of the country. Instead, the available space based technologies are now being incorporated towards militarisation of Indian Space Programme and in near future, possibly towards weaponisation as well. Now that the Chinese have already carried out the ASAT test in 2007 and unfortunately thereafter their approach towards India is becoming more and more belligerent, hence an appropriate response to the Chinese ASAT should be the only choice. Thus it is imperative for India to develop a counter ASAT to deter any Chinese mis-adventure towards the Indian space assets, as the country is getting more and more dependent on satellites for its economic development. It would therefore strategically be prudent for India to secure its space assets from Chinese threat. Towards this, an Indian ASAT perhaps will be the most potent and credible option for India, which can create a desired deterrence for China, in the form of a counter weapon in space.

The direct-ascent anti-satellite (ASAT) weapon that China tested in January 2007 appears to be part of larger efforts to develop a range of ASAT capabilities, including ground-based lasers and jammers. When deployed, China's direct-ascent ASAT could hold Indian satellites in low earth orbit (LEO) at risk. Other potential Chinese ASAT capabilities might be able to disrupt the use of satellites in higher orbits, including the Global Positioning System and India's IRNSS. Loss of these space assets would significantly affect the Indian space capabilities, both economically and militarily.

The Economic Impact

We all appreciate that using space assets has become an everyday event, much as television has over the past fifty years. When we turn on the TV, we simply expect the picture and sound to be there; no one speaks with awe about how the video and audio waves appear. Many people start their day by driving to work in an vehicle with a GPS enabled road navigation system that depicts their present location and guides them to their desired destination; directs them across town following instructions to a predetermined destination; stop to gas-up by using a credit card at the pump; and remove money from an automated bank teller machine from their account that could be from a different bank in another part of the country. They will think nothing about the technological wizardry, but these set of transactions, location, directions, link to credit card and banking accounts are all made possible by instantaneous access to multiple satellite constellations, something we all take for granted. The satellite systems can also provide navigation for civilian airliners, identify underground water in water scarce regions of the country, and mark the destruction of the forests, in addition to numerous other everyday services we have all come to expect from a modern society. The failure of a single satellite in May 1998 disabled 80% of the pagers in the United States, as well as video feeds for cable and broadcast transmission, credit card authorization networks, and corporate communication systems. If the Global Positioning System (GPS), a multi-satellite constellation originally designed for military navigational assistance were to experience a major failure, it would disrupt air traffic operations around the world; cripple the global financial and banking system; and could in the future threaten air traffic control. Space, therefore whether we realize it or not, plays an increasingly important role in everyday life. Thus, loss of a satellite as a result of enemy attack through an ASAT would cripple the country's economic activities to an un-imaginable level.

The Strategic Impact

The Chief of General Staff of the Chinese People's Liberation Army, General Fu Quanyou highlighted on 20 Feb 1999, the PLA's thinking on the conduct of future war and the preparations for the same by opining

> *"The space of operations becomes all-dimensional and military action will be conducted simultaneously in many fields, on land, in the air, in outer space and in the electronic field."*

Most strategists world over believe that ASAT weapons are one of a series of asymmetric capabilities that China is developing, primarily to exploit the US military vulnerabilities. This perception emanates from the fact that the US military is heavily dependent on its space based assets for secure communication, navigation, weather, precision bombing and co-ordinated military operations across the globe and the Chinese ASAT can potentially exploit this vulnerability and reduce the ability of potential adversaries to operate with freedom in the Indo- Pacific theatre. One space expert argued that ASAT weapons are a logical and relatively inexpensive response to the US military dominance, which rests heavily on space capabilities. The 'Pacific Pivot' propounded by President Obama in early 2012 and the consequent relocation of military forces to the Indo Pacific theatre has propelled the Chinese space developments and deployments. This is apparently visible from the fact that the Chinese have surpassed the US in launching of satellites in year 2012 by 19:18 and became the second nation after Russia, to launch maximum number of satellites for economic and of course military utilization.[2]

---

2 Doug Messier in Parabolic Arc, "China Surpassed U.S. in Launches, Payloads in 2012 " on 02 Feb 2013 accessed through http://www.parabolicarc.com/2013/02/02/china-surpassed-u-s-in-launches-payloads-in-2012/

Though the Chinese ASAT capabilities are a threat to US space assets, it complicates the strategic military relationship between India and China as well. The Chinese forces may attempt to temporarily blind reconnaissance and remote sensing capabilities through lasers, while jamming communication links. GPS signals could disrupt navigation and (more importantly) precision targeting. These efforts could be coupled with cyber-attacks to disrupt and delay the response of the US armed forces. This strategy could be conducted in whole or in part, without complete integration of systems, thereby severely impacting the Indian security and its counter military response.

The Build-up for an Indian ASAT

The Chinese ASAT test of 11 Jan 2007 opened a new flank of vulnerability in India's $12 billion (Rs 60,000 crore) space infrastructure.[3] India has a large constellation of satellites, which include the latest, the sophisticated and high value Radar Imaging Satellite (RISAT-I), which can identify one-metre wide objects on earth from space. However, this satellite can be rendered unusable if China happens to use KKE ASAT similar to the one it demonstrated on 11 Jan 2007. China's alarming test spurred India's quest for a similar counter strike capability in the form of satellite-killing system.[4] This will give India a strategic balance in the form of deterrence, which can be used against China, in the event of any celestial shoot - out in space between the two Asian neighbours, who have long drawn territorial dispute which can raise the heat leading to a war. A glimpse of the same was visible in the Depsang bulge in Ladakh on 15 Apr 2013, where the Chinese troops intrude 19 Km inside the Indian Territory and pitched tents[5] which they refused to dismantle even after few rounds of border meetings.

India took cognizance of the need for a counter ASAT

---

3 BharathGopalaswamy and Harsh Pant, "Does India need anti-satellite capability?" Rediff News, Feb. 9, 2010

4 SandeepUnnithan in India Today.in on April 28, 2012 'India attains the capability to target, destroy space satellites in orbit' accessed through http://indiatoday.intoday.in/story/agni-v-launch-india-takes-on-china-drdo-vijay-saraswat/1/186367.html

5 N C Bipindra in The New Indian Express on 25 Apr 2013 "Fresh Chinese incursion in Ladakh area" accessed through http://newindianexpress.com/nation/Fresh-Chinese-incursion-in-Ladakh-area/2013/04/25/article1560071.ece

capability; however there is no governmental programme which has been set for development and testing of an ASAT. In this regard Dr VK Saraswat, the Scientific Advisor to India's Defence Minister, said in Jan 2010, "India is putting together building blocks of technology that could be used to neutralize enemy satellites. These are deterrence technologies and quite certainly many of these technologies will not be used." [6] Within a month in February 2010, he was quoted by Indian News Agency PTI, "India will validate the anti-satellite capability on the ground through simulation, as there is no program to do a direct hit to the satellite." [7] Dr Saraswat has conveyed to the media time and again that India does not have a policy to attack anybody in space as it doesn't have a formal anti-satellite weapon policy of attacking satellites in space. But as part of the Ballistic Missile Defence Program, India does possess all the technology elements required to integrate a system through which we can defend our satellites or take care of future requirements. [8] He has amplified in clear terms that "Space security involved a gamut of capabilities including the protection of satellites, communication and navigation systems and denying the enemy the use of his own space systems. These technologies would be developed as part of the country's totally indigenous Ballistic Missile Defence Programme."[9]

Thus an opinion was slowly shaping up within the military and strategic community for an Indian ASAT, as credible deterrence against the Chinese ASAT. In order to give shape to this perceived threat to the Indian space assets and co-ordinate the development of space capabilities towards military use, a Space Security Coordination Group (SSCG) was set up in 2010. This was chaired by the National Security Adviser and involved representatives of DRDO, Indian Air Force (IAF) and National Technical Research Organisation (NTRO). Besides laying down the Government's space policy, this body will

---

6 Thiruvananthapuram, SagarKulkarni, "India readying weapon to destroy enemy satellites: Saraswat," Press Trust of India, Jan. 3, 2010

7 "India has anti-satellite capability: Saraswat," Press Trust of India, Feb. 10, 2010

8 "India has technology to defend satellites: top defence scientist," Xinhua General News Service, Feb. 12, 2011

9 S. Paneerselvam and P. Soma, "Anti-satellite Weapons (ASAT): A Status Review and Perception for an Indian ASAT," presentation given at "Space, Science, and Security: The Role of Regional Discussion," New Delhi, India, Jan. 19–21, 2011

also coordinate response on an international code of conduct in space and other policies on space and space laws.

The Success of AGNI – V: A Game Changer

The successful trial of 5,500 km Agni- V Intermediate Range Ballistic Missile on 19 Apr 2012, which Dr Saraswat calls a "game changer", is another step towards the Indian capability to target objects in space. The Agni-V brought India supposedly very close to having achieved the ASAT technology. As per the top defence scientists, India now has the capability to target and destroy space satellites in orbit. "Today, we have developed all the building blocks for an anti-satellite (ASAT) capability," remarked Scientific Adviser to the Defence Minister and DRDO Chief Dr VK Saraswat in an interview with India Today.[10] This capability once validated and deployed will give India a credible deterrence against China. In this context talking to India Today quoted a top government source, who does not want to be named, brought out that DRDO will field a full-fledged ASAT weapon based on Agni and AD-2 ballistic missile interceptor by 2014.[11] This was also confirmed by Dr Saraswat who said that

> *"India will not test this capability through the destruction of a satellite. Such a test risked showering lethal debris in space that could damage existing satellites. Instead, India's ASAT capability would be fine-tuned through simulated electronic tests".*[12]

The first test of a PDV interceptor validated another key ASAT milestone capability, wherein this slender two-stage missile can destroy incoming ballistic missiles at an altitude of 150 km. This Ballistic Missile Defence (BMsD) which will is being developed to protect the country from hostile ballistic missiles, has developed the critical elements required to destroy satellites, if it is utilised in an ASAT role. The DRDO's Long Range Tracking Radar can scan

---

10 Sandeep Unnithan in India Today.in on 28 Apr 2012 'India attains the capability to target, destroy space satellites in orbit' accessed through http://indiatoday.intoday.in/story/agni-v-launch-india-takes-on-china-drdo-vijay-saraswat/1/186367.html

11 Ibid

12 Ibid

targets over 600 km away. The 'kill vehicle' has been developed as part of the ballistic missile system. It has both electronic and radio-frequency guidance that can home in on ballistic missiles and satellites. Unlike a ballistic missile, a satellite has a predictive path. A satellite has a diameter of one meter while the BMD system can track and destroy targets less than 1 metre. Since all the technologies required to develop an ASAT weapon has already been validated in the successful testing of Agni-V and Indian BMD, India can now go ahead with an ASAT technology demonstrator.

The Relevance of an Indian ASAT

India is a developing nation with a major role in Asia. Contrary to the global trend, the pace of its development is moving ahead with positive growth, even in the times of recession world over. India's growing stature in world is reflective of the new role which the country has started to play in the world affairs. This economic growth has resulted in a bigger geo-political and geo-strategic role for India, not only in Asian continent, but also in the World Order which is apparently transforming into a New Economic World Order. In this new scheme of things, the space assets have been instrumental in shaping the Indian success story. As a result more and more satellites are being launched to support large number of services which are poised to shore up the economic activities of the country, apart from supporting a large number of social welfare programmes.

Offence is the best defence. Apart from military personnel and military strategists, even a common man understands this axiom. However when extrapolated to the national level, it is surprising that the "killer instinct", which pumps the adrenaline and justifies one's belief in "offence is best defence", appears to be amiss. Perhaps that's the only justification for the country's political leadership to withhold the decision of developing its own ASAT programme, inspite of the fact that the country, as per the top defence scientists, does have the capability of developing an ASAT weapon for India, as and when the situation demands. In this regards Dr VK Saraswat, the Director General of Defence Research and Development Organisation has gone on record while briefing the reporters on the occasion of 97th Indian Science Congress at Thiruvananthapuram in Jan 2010, "India is putting together building blocks of technology that could be used to neutralize enemy satellites. We are working to ensure space security and protect our satellites. At the same time we are also working on how to deny the enemy access to its space assets."[13] The nation's determination towards building as ASAT for India was also visible in the statement of Dr Saraswat after the successful test of AAD missile programme in Apr 2012, wherein he said,

> *"Agni-V's launch has ushered in fantastic opportunities in building ASAT weapons and launching nano/micro satellites on demand. The said ASAT would include marrying Agni-V's propulsion system with the kill vehicle (AAD) of the successful II-tier BMD system. Since the ASAT is required to reach an altitude of 800 km altitude, Agni-V will give the boosting capability and the 'kill vehicle', with advance seekers, will be able to home into the target satellite. India will not test ASAT capability through destruction of a satellite so as to avoid space debris. Instead, the Indian ASAT capability would be fine-tuned through simulated electronic tests".[14]*

Though the Chinese ASAT has spurred the quest for an Indian ASAT capability, but there appears to be an intense debate going on between the Indian space and the strategic community on two counts: firstly, the necessity of possessing an ASAT, and secondly its impact on the Security Architecture in the Asia Pacific region. India appears to be sceptic about disturbing the Asian security architecture, which is led by China. Yet, it also appreciates the vulnerability of its space assets and thus the necessity to protect them in view of their importance as national assets, which are indeed the engines for economic growth and development. Since there is so much at stake for the country, the Indian satellites need to be protected from any Chinese mis-adventure in future. Dr Kasturirangan did appreciate this necessity immediately after the Chinese tested their ASAT on 11 Jan 2007. Reacting to the Chinese ASAT test, which took the

---

13 Sagar Kulkarni from PTI Thiruvananthapuram on on 03 Jan 2010,"India readying weapon to destroy enemy satellites: Saraswat,"

14 "India developing anti-satellite weapons" in Space War, a web portal on space weapons and technology on 23 Apr 2012, accessed through http://www.spacewar.com/reports/India_developing_anti-satellite_weapons_999.html

world including India by surprise, Dr Kasturirangan told PTI in an interview

> *"India has spent a huge sum to develop its capabilities and place assets in space. Hence, it becomes necessary to protect them from adversaries. There is a need to look at means of securing these."*[15]

With this background, the developments efforts in respect of the Indian ASAT moved forward. In a televised press briefing during the 97th Indian Science Congress in Thiruvananthapuram, the DRDO Director General VK Saraswat announced that "India was developing lasers and an exo-atmospheric kill vehicle that could be combined to produce a weapon to destroy enemy satellites in orbit. The kill vehicle, which is needed for intercepting the satellite, needs to be developed, and that work is going on as part of the ballistic missile defence program".[16]

Highlighting the governments stand on the ASAT programme, Dr Saraswat said that the propulsion module and kill vehicle already existed in principle on the Agni (missile) series of ballistic missiles, but there is no formal 'anti-satellite weapon project' in the offing as yet. He however indicated that the anti-satellite weapons could be developed as part of the Indian Ballistic Missile Defence Program, which will complete the development stage in totality by 2014.[17] India's quest for a deterrence against China has also been echoed by Rajat Pandit, a noted defence journalist who brought out that India had identified development of ASAT weapons "for electronic or physical destruction of satellites in both LEO (2,000-km altitude above earth's surface) and the higher GEO-synchronous orbits" as a thrust area in its long-term integrated perspective plan (2012–2027).[18]

---

15 Press Trust of India, 14 Sep 2009 "Ex-ISRO chief calls China's A-SAT a cause for worry,"

16 Peter B. de Selding, 11 Jan 2010, "India Developing Anti-Satellite Spacecraft" accessed through http://www.space.com/7764-india-developing-anti-satellite-spacecraft.html

17 "India developing weapon system to neutralize enemy satellites".Xinhua News Agency. 2010-01-03.

18 Pandit, Rajat (May 25, 2010). "India to gear up for 'star wars'".The Times Of India.

However, the silence regarding an ASAT programme maintained by India after the Chinese ASAT test in 2007, could apparently be for the lack of technical competence for developing a potent ASAT weapon or due international considerations. Since Dr VK Saraswat has already gone on record that the technology and capability for an Indian ASAT is available within the country, there could be two pragmatic questions bothering the Indian policy makers. The two basic questions which India needs to consider prior to initiating an ASAT programme are:

- Should India disturb the status quo in Asia by testing its own ASAT weapon?

- International fallout of ASAT test by India, specially the Chinese reaction.

Indian Geo-strategic Location in the Asian Pacific Region

Before understanding the necessity for India to go ahead with its ASAT programme, it is pertinent to appreciate the peculiar geo-strategic location of India in the Asia Pacific region and the changing dynamics in India's immediate neighbourhood. India occupies a very strategic position in Asia Pacific in general and South Asia in particular. The Indian Ocean is the hub for intense international activity, both for trade and commerce and also for strategic and military considerations. As a result of the hotspots in Asia, which erupts with regularity across the length and breadth of the Asian continent from the Gulf of Hormuz to Malacca Straits, any event taking place in such a vast geographical expanse has a direct impact in Indian security architecture; it could be purely a strategic threat from a military perspective or economic threat to trade and commerce.

So inter-woven are the issues in this hyper sensitive region that even a small event has a potential to impact the Indian economic and strategic environment. The Indian Exclusive Economic Zone (EEZ) in the Arabian Sea and the Bay of Bengal and areas immediately beyond Indian EEZ, witness intense movement of trade and commerce through sea. 90 % of India's international trade by volume and 77 % by value are carried by this sea-lane.[19] India`s strategic position in the Indian Ocean is very apparent, when seen in the context of the massive flow of trade through the sea-lines of communications, which join the major regions of West and East. Seaborne trade passing through Indian Ocean amounts to almost 15% of entire world trade. Nearly 200 ships cross Malacca strait every day. Of this 28% passes through Strait of Hormuz, carrying nearly 20% of the world trade in volume. Over half of United States requirement of oil passes through this route. One-third of the total ships in the world and over half of the entire world shipping capacity take passage through these choke points.[20]

Though this busiest sea lane impacting the world trade is governed by the UNCLOS, but there are violations, inadvertently and at times intentional, which raises the military and the diplomatic humbug to an alarming level. The recent killing of Indian fishermen off the coastal state of Kerala by the Italian Marines onboard Italian Merchant ship is a case in point. Similarly, Chinese refusal to acknowledge UNSLOC in the South China Sea for Philippines is another example of how issues grow and lead to a potential trial of strength between two maritime neighbours. Much nearer home, the disputes over fishing rights between India-Pakistan and India-Sri Lanka and the resultant killings and arrests of fishermen are regular events, which persistently keep tickling the fragile relations between neighbours. Moving further to the East, the South China Sea is witnessing intense military activity and diplomatic engagements between China, Philippines and Vietnam over the Spartlys Island. Similarly, China and Japan are locked in a confrontation over the Diaoyu Islands, known as Senkaku in Japan, which triggered street protests across China and saw attacks on Japanese businesses in China.[21] Besides this, Japan and South Korea are entangled in a diplomatic row over the possession of Dokdo Island in South Korea – also called Takeshima islands in Japanese.

---

19 "Indian Maritime Trade, Geography of India" accessed through http://www.indianetzone.com/37/indian_maritime_trade_geography_india.htm

20 ibid

21 "China, Japan in tense talks on disputed islands" published on 27 Sep 2012 accessed through http://www.thenews.com.pk/Todays-News-1-134280-China-Japan-in-tense-talks-on-disputed-islands

Besides the busy sea lane close to its EEZ, India has another unique distinction of being located in the worst hotspots of the world. On one side is a hostile neighbourhood with whom India has inherited animosity since its independence and has already fought three full-fledged wars and an undeclared war in Kargil. Even now, the border skirmishes are routine affairs. On the other side is China, which waged a war against India in 1962, when India was propagating and practicing the 'Principles of Panchsheel'. Even now, because of its intrusive policies towards India, China is still viewed with suspicion. China continues with the illegal occupation of Indian Territory of Aksai Chin in Jammu & Kashmir. China's 'Shifting Core Interest' is also a cause of concern for the Indian Establishment, wherein China has also laid claims over the state of Arunachal Pradesh apart from the disputed land borders adjoining Ladakh, Himachal and Uttarakhand. As a result, border intrusions by Chinese troops are a regular affair. While there have been no major border confrontations with China since 1962, nor any actual skirmish between the two armies, India has recorded as many as 500 "transgressions" by Chinese troops across the LAC in all the three sectors – Western (Ladakh), middle (Uttrakhand and Himachal) and Eastern (Sikkim and Arunachal) – just since January 2010 as reported by a leading Indian Newspaper, The Times of India.[22] However, the border intrusion by Chinese PLA, 19 Km deep inside the Indian territory in the Depsang area of Ladakh on 05 May 13, escalated to a new level, wherein the Chinese Prime Ministerial visit to India was almost cancelled. There have also been reports of smuggling of arms and ammunitions from the Chinese borders and also tacit support to separatist organisation, which is a cause of concern for India. On strategic level too, the encirclement of India by creating ports and facilities in the Indian Ocean RIM countries for berthing of Chinese ships both commercial and naval, is another issue bothering the Indian strategists.

Apart from the big neighbours in West and North, the relations with other smaller neighbours like Sri Lanka, Maldives, Myanmar, Nepal and Bangladesh has been witnessing lots of ups and down. Though India has done a lot to dispel the myth of "Big Brother Syndrome" with which these neighbours are stuck with, yet the

---

22 "China needles India in Eastern Sector" reported by Times of India Sep19, 2012

relations are moving "one step forward and two steps back" because of some issues or the other. As a result, the SAARC region as a whole has been deprived of economic development and the well-being of huge population residing in these countries. The mistrust amongst these countries has held them captive to misconceived notions and intentions about each other, which is for sure affecting collective growth of one of the poorest region in the world. Hence, as a result of massive influx of refugees from Bangladesh, Nepal and Sri Lanka, India's national security is getting compromised.

The Security Scenario in South Asia

India, perhaps is the only country in the world, grappling with "a three front strategic challenge"

- Pakistan

- China

- Drifting SAARC nations

Each of these nations has been associated with India right from its independence. Even though they had a common past and share the legacy of history, unfortunately their differences could not be settled even after 65 years of independence. Instead, the irony is that the region is still mired with confrontations, both on military and economic fronts. This perhaps has been the single reason for withholding the growth and development of the Indian sub-continent as a whole. As a result, the Indian sub-continent still continues to remains one of the poorest regions in the world and there are no signs of improvement in the near future, unless some divine intervention takes place, which for sure would be a miracle of the 21st century.

The lack of collective economic growth of the SAARC region is a crucial factor for huge refugee influx to India. This is not only creating law and order problems all across India, but has become a serious risk to the national security and the delicate and the fragile social fabric of the country. The recent riot in Assam is one such indicator of things to come, if this issue is not addressed in all its sincerity. Though enough has been written about the threat emanating from the Indian neighbourhoods, it would be prudent to highlight these threats, which are not only needling the Indian security architecture, but are economically bleeding the Indian State. Few of these issues are highlighted below

- Sponsored terrorism from across the Indian borders

- Injection of fake currency

- Money laundering

- Drug trafficking

- Human Trafficking

- Illegal migration

- Border disputes

- Disputes over river water sharing

- Disputes over fishing rights

- Fermenting and stroking of religious fundamentalism

- Smuggling of arms and ammunitions

Indian Growth and Economic Rise

It is indeed a remarkable achievement for the Indian state to withstand all burning issues and yet manage to secure economic development. The country has posting around 6 % overall growth rate in last two years, when world on the whole has been facing severe recession. It has also developed ways and means to address its threat perception and has been successful in managing and redressing few of them as a result of its astute resolve and diplomatic initiatives. Inspite of these positive developments, the Indian state is grappling with vulnerabilities, which have a profound impact on its economy. However, these vulnerabilities have been tackled well and gave the country a chance to consolidate its position in world affairs based on its new found economic status. This economic resurgence has ensured that India takes a proud place next to China in leading the 'Asian Growth Story' against all odds and in testing times, when the world is going through a phase of recession and economic crisis. This economic strength acquired by the Indian state has been instrumental in placing it very high in the evolving 'New Economic World Order'. One of the most significant aspects of the Indian success story is the role played by technology and amongst them; the "satellite based technologies" contributed heavily, thereby heralding a new era.

The satellites, apart from providing the services towards the societal good, played a catalytic role in giving a fillip to the economic activities. Thus the satellites became the 'engines of economic growth' in the 21st century. The strong support which satellite based services provided to trade and commerce, ensured rapid development and consolidation of economic activities. Coupled with the abundance of "soft skills" available in the country, the satellite based services became all pervasive. As a result India was able to rewrite its success story and lead a path for of economic transformation, resulting to an unprecedented economic growth.

The growth of space technology and its impact on human activities have gone to prove that man has never been so dependent on one single asset than he is today on Satellites. India has also managed its satellite development plans very well. It was able to successfully develop an indigenous satellite manufacturing capability and equally robust launch capability. The efforts put in by the Indian scientific community; especially the space scientists are creditable. Today, India is ranked fifth space power after the US, Russia, China and EU. It can boast of indigenously manufacturing and launching a whole range of satellites for

- Communication

- Navigation

- Weather

- Land mapping

- Surveillance

The Necessity and Urgency of ASAT

The Necessity

Since the country's economic development is so dependent on satellite based technologies, in the face of existing security scenario and the development and testing of ASAT by China, the Indian space assets need to be secured from the physical damage and destruction. This destruction could be from internal threat of sabotage and physical damage of ground support systems or threat of physical destruction of the satellites, which are vulnerable to ASAT threat from our adversaries. Towards this, India definitely needs to develop its ASAT capability. The technology apparently is available within the country, as has been claimed by the defence scientists from DRDO. If that be the case, the time has come for India to initiate its effort towards demonstrating its ASAT capability, which would act as deterrence to our adversaries from initiating any attack on Indian space assets. The physical demonstration of ASAT capability is in the interest of the country, as "deterrence is potent and credible only when it is physically demonstrated".

The Urgency

The necessity of testing a potent ASAT weapon to fulfil the Indian strategic requirement has been well appreciated in the strategic community. They also have appreciated that the strength required to negotiate and talk peace with powerful neighbour like China will come from a potent deterrence in the form of ASAT, which can effectively provide an asymmetric advantage to India against China. Besides this, the defence scientist have already assured that the technology for building an ASAT weapon does exist within the country, only the political go ahead is required to build the same. However, the only factor needs to be appreciated is the international reaction towards testing of an ASAT weapon.

A 'Code of Conduct' for safe practises in the outer space was proposed by EU in 2008 and resubmitted to the UN in 2012, after two rounds of consultations and review. The current Draft Code of Conduct lists as its main purposes:

- Enhancing the security, safety and sustainability of all outer space activities.

- Endorsing best practices.

- Supporting existing international space law, such as the Outer Space Treaty (OST) of 1967.

Once the CoC is ratified by the UN, the following guiding principles are likely to be enforced on the space faring nations:

- Freedom of access to space for peaceful purposes without interference and with respect for the long term sustainability and safe conduct of space activities.

- The inherent right of all States to collective and individual self defence.

- The responsibility of States to avoid harmful interference in space.

- The responsibility of all States, in their pursuit of space activities, to promote peaceful uses and avoid conflict in space.

After prolonged deliberations the CoC has now been taken up at the UN for debate a 'Group of Government Experts'. Since there is a rising demand to check proliferation of space weapon so as to check the menace of space debris, there is a strong possibility that the CoC is ratified and adopted by the UN in 2013-14. If so, then the CoC could be a pre-cursor for an 'ASAT ban treaty' which perhaps will be initiated by the 'Big-3 Space Player,' the US, Russia and China. This ASAT ban treaty will be a restrictive regime on the lines of NPT, so as to put a ban on further testing of ASAT weapons by any state. If this restrictive regime is ratified under the aegis of the UN, there would be a total ban on any further testing of ASAT weapons and India may lose a chance to test its desired ASAT capability.

India, therefore needs to take a judicious and practical approach towards developing and testing of its ASAT capability on priority or else it will lose out in the strategic counter balance against Chinese ASAT. As regards the international fall out and sanctions consequent to a possible ASAT testing by India, the issues and way out is discussed in successive paragraphs.

Fallouts of a Possible ASAT Test by India

If India decides to develop its ASAT capability and carry out a test, a strong international reaction is apparent. Thus, India needs to prepare in advance the ways and means to tackle the fallouts, which are likely to impact the country's economic development. Since the international community is paranoid about the space debris, the repercussions are likely to be strong and thus India needs to guard against the negative impact of testing an ASAT. Since restrictive regimes or negative list is nothing new for India and also the fact that India has experienced to live with nuclear apartheid, the expected restrictive regimes consequent to ASAT testing by India, is not bound to have significant impact. The present day India is economically much stronger to handle the impact of any regime and would be able to handle the fallouts, with little effect on its economic activities and international relations.

However, there is a way to avoid the international fallout over ASAT test, if the space debris are avoided or controlled in such a way that it re-enters the earth's atmosphere and automatically burn out. This can be conducted on the lines of the US ASAT test on 21 Feb 2008, wherein the US destroyed its malfunctioning spy satellite USA-193 using a RIM -161 Standard Missile 3. We need to appreciate the whole gamut of this supposed KKV Test by the US. The satellite, USA-193 was an American spy satellite, which was launched on 14 December 2006 by a Delta II rocket, from Vandenberg Air Force Base. It was reported about a month after launch that the satellite had failed. In January 2008, it was noted that the satellite was decaying from orbit at a rate of 1,640 feet (500 m) per day.[23] On 14 February 2008, it was reported that the US Navy had been instructed to fire an SM-3ABM weapon at it, to act as an anti-satellite weapon.[24] The **RIM-161 Standard Missile 3 (SM-3)** is a ship-based missile system used by the US Navy to intercept short- to intermediate-range

---

23 "U.S. plans for falling satellite – CNN.com" accessed through Wikipedia http://en.wikipedia.org/wiki/Anti-satellite_weapon

24 Associated Press – " Broken Satellite Will Be Shot Down" accessed through Wikipedia http://en.wikipedia.org/wiki/Anti-satellite_weapon

ballistic missiles as a part of Aegis Ballistic Missile Defence System.[25] Although primarily designed as an anti-ballistic missile, the SM-3 has also been employed in an anti-satellite capacity against a satellite at the lower end of Low Earth Orbit.[26]

The primary reason for destroying the satellite, according to the US Government, was the approximately 1,000 lb (450 kg) of toxic hydrazine fuel contained on board the satellite, which could pose health risks to persons in the immediate vicinity of the crash site, should any significant amount survive the re-entry.[27] On February 20, 2008, it was announced that the launch was carried out successfully and an explosion was observed consistent with the destruction of the hydrazine fuel tank.[28] Experts debated whether the hydrazine tank would have survived an uncontrolled re-entry. However, if it had, any human fatality would still have been very unlikely. Although hydrazine is toxic, a small dose would not have been immediately lethal. The chance of the (assumed) hydrazine tank landing close enough to at least one person for that person to be killed if he or she lingered in the vicinity of the crash site was about 1 %, while the cost of the intercept was about $100 millions.

The intercept, however, was widely interpreted as a demonstration of US capabilities in response to the Chinese anti-satellite test a year earlier. The intercept was different from typical ASAT missions, in that it took place at a much lower altitude (133 nautical miles or 247 Kms) than would normally be the case, and the SM-3 missile as currently deployed would not have adequate range and altitude reach for typical ASAT missions in low-Earth orbit. However, the warhead was shown capable of hitting a satellite at orbital closing speeds. While an SM-3 missile would require significant modification to fill an anti-satellite role, the test was a proof of concept, demonstrating that it can operate in such a role if required. The best part of this demonstration was that it can be

---

25 Raytheon Completes SM-3 Test Flight Against Intermediate Range Ballistic Missile, Raytheon Company, Retrieved 6 September 2011

26 Pentagon news briefing of February 14, 2008 (video, transcript): although no name for the satellite is given, the launch date of 2006-12-14 is stated.

27 "Navy missile hits dying spy satellite, says Pentagon – CNN.com". February 21, 2008.

28 "US shoots down toxic satellite". Daily Telegraph (Australia). 2008-02-20.

achieved with 'almost negligible debris'; hence not much of hue and cry can be expected from other space faring nations, except from the proponents of weapon free space.

Chinese ASAT Weapons and the Comprehensive Indian Response

It is a well-known strategy to 'fight threat with threat'. The Cold War adversaries know it much better, as they eventually countered the nuclear threat from each other by espousing the MAD (Mutually Assured Destruction) strategy. It indeed was a powerful deterrence, which was successful to ward of a nuclear attack by the Cold War adversaries, even during the peak of the Cold War. Similarly, India should draw a parallel from the Cold War strategy and develop, demonstrate and deploy its own ASAT weapon as a strategic deterrence to China from using ASAT weapon against Indian space assets.

Keeping in view the existing geo-political scenario and the emerging belligerent Chinese postures and strong arm tactics by China, India therefore needs to undertake the following:-

- Militarisation of space for effective C4ISR, which will strengthen, not only the defence preparedness of the country but will also assist in operational planning.

- Develop a reliable 'Space Situational Awareness' architecture to identify threats in space.

- Safeguard for space assets by developing requisite defensive counter space technologies.

- Weaponisation of space for offensive space operations to deter adversaries from initiating attack against own space assets.

- Dialogue with the UN and other nations to ensure strong lobbying for early ratification of Code of Conduct for space activities in the outer space and thereafter formulation of an appropriate treaty for a new regime of "Weapon Free Space".

Apart from this, the country should also plan and develop ways and means for the following options, in order to counter the potential Chinese ASAT capabilities

- Potent and Credible Deterrence.

- Direct attacks against Chinese ASAT systems.

- Space-based weapons to attack Chinese ASAT systems or space assets.

- Co-orbital Satellites that are harder to find and harder to hit.

- Constellations of small satellites for ASAT operations.

In addition to the purely tactical military responses referred above, India must also work on non-military means to counter Chinese attempts towards militarisation and weaponisation of space. Few options are highlighted below:

- Learn to fight without satellites.

- Consider diplomatic solutions.

- Adopt an international code of conduct on space behaviour.

The Way Ahead

India now has the capability to target and destroy space satellites in orbit. This capability, once demonstrated and proven, will give India, deterrence against China. Besides the BMD system and an ASAT capability, India needs to develop the following capabilities to deter to the adversaries:

- Space Surveillance Network Satellites equipped with radars which will

    - Ensure launch detects, tracks, measures and catalogue the orbiting objects and space debris

    - Improve accuracy

    - Better timeliness

- Increased surveillance coverage

- Urgent need for a dedicated National Space Command to co-ordinate and put in place an effective counter space operation apparatus in place.

- Space assets with military applications need to be placed under National Space Command.

- A dedicated Space force to be raised from within the armed forces cadre.

- Dedicated training of space forces to be initiated for effective and efficient adaption of space technologies in the armed forces.

- India's KKE ASAT to be demonstrated for strategic balance in Asia (on fast track).

- Space assets with military applications to be concurrently developed.

- Integration of military applications of space in armed forces.

If these recommendations are finalized and incorporated into the Indian strategic buildup, they will be able to build a comprehensive "Indian Space Security Architecture" and India's offensive space capabilities will form the crucial component, which will ensure, **"a secure aerospace for India to propel its economic activities in the 21st century, thereby making it a leading world power."**

PART – II

INDIAN BALLISTIC MISSILE Defence (BMD) SYSTEM

The DRDO is developing a layered ballistic missile defence system which will be developed in two phases and is likely to be ready for deployment by 2015. DRDO Chief Dr VK Saraswat confirmed this while addressing the media at Thiruvananthapuram on 21 Mar 2010, where he told reporters that, "the Ballistic Missile System (BMD) is being developed in two phases under a capability based deployment plan".[1]

- **Phase I.** In the first phase, which is currently underway, DRDO will develop and deploy a system for defence against missiles with less than 2,000 Km range like Pakistan's Ghauri and Shaheen missiles and China's solid-fuel Dongfeng-21 (NATO designation: CSS-5).

- **Phase II.** In the second phase, the system capability will be upgraded to defend against missiles with ranges greater than 2,000 Km that and those which can additionally manoeuvre and deploy decoys. However, the system in this phase will require longer range radars (Detection range of 1,500Km as opposed to 600 Km for Phase 1 radars), and new hypersonic interceptor missiles flying at Mach 6-7 (As opposed to Mach 4-5 for Phase 1 missiles) with agility and the capability to discriminate against BMD counter measures like decoys and chaffs.

The Indian BMD System: An Overview

The Indian BMD system will be based on radar technology for tracking and fire control which the DRDO has developed jointly with Israel and France respectively. It will be implemented as a II-

---

1 "Ballistic Missile Defence (BMD) System" , IDP Sentinel Accessed through http://idp. justthe80.com/missiles/ballistic-missile-defence-bmd-system

tiered, terminal phase interceptor system comprising of

- Prithvi Air Defence (PAD) exo-atmospheric interceptor missile for intercepting targets outside the atmosphere, up to an altitude of 80 Km.

- Advanced Air Defence (AAD) endo-atmospheric interceptor missile for intercepting targets up to an attitude of 30 Km.

- 'Swordfish' Long Range Tracking Radar (LRTR). This radar has been developed jointly by LRDE, Bengaluru and ELTA of Israel. It is based on the Israeli Green Pine early warning and fire control radar imported by India from Israel in 2001-2002.

- Multi-Functional Guidance radar that tracks the incoming missile in its terminal phase and guides the interceptor missile onto the target. The DRDO developed the guidance radar in collaboration with Thales, France.

- The integrated exo and endo-atmospheric systems will be able to offer a hit-to-kill probability of 99.8 %

Prithvi Air Defence (PAD) Missile

PAD is a two stage missile with a maximum interception altitude of 80 Km. It has a solid fuel first stage and a liquid fuel second stage, both of them designed by DRDO. The missile uses inertial navigation system with mid-course correction from LRTR. In its terminal phase, it switches to active radar homing to reach the target for a kill. It uses a 'manoeuvring gimballed directional warhead' which can rotate 360 degrees to explode towards the incoming missile in order to destroy it. DRDO claims to have successfully tested this niche technology, which till now has only been carried out by US and Russia. Since it is a directional system, the 30 kg warhead is able to generate an impact equivalent to a 150 Kg Omni-directional warhead.

Prithvi Defence Vehicle (PDV) Interceptor Missile

The PAD missile will be replaced with another indigenous missile system, the PDV, which has two solid fuel stages. It will be capable of intercepting enemy missiles at altitudes up to 150 Km. Like the PAD missile, it has been designed by DRDO and will feature a directional warhead. This system will be equipped with an innovative system to allow the missile to manoeuvre at altitudes up to 150 Km, well outside the earth's atmosphere. The PDV interceptor missile is planned to be the mainstay of the defence shield.

Advanced Air Defence (AAD) Interceptor Missile

The endo-atmospheric interceptor AAD is a 7.5m long, single stage solid fuel missile, which is equipped with ring laser gyro based inertial navigation system, a hi-tech computer and electro-mechanical actuators. This missile system is totally under command by the data up-linked from the sophisticated ground based radars to the interceptor. In an interview with the Indian Express in April 2012, comparing the AAD missile with Patriot-III of the US, DRDO Chief Dr VK Saraswat informed that AAD missile system is equivalent to Patriot-III, which uses an active guidance. This system with active guidance scores much better over the semi-active guidance used in Patriot-II. Highlighting the capabilities of AAD system, Dr Saraswat said "The philosophy in this case is first you launch the missile based upon the data about the target collected by a ground radar. Once the missile goes close to the target, the homing seeker homes onto the target and starts tracking autonomously." [2]

Projectile Charge Interceptor Warhead (P-Charge)

The AAD interceptor is equipped with a P-charge warhead that can penetrate thick armour of steel and cause damage with a high hit (repeat hit) density. As per DRDO scientists this means that the number of holes the P-charge can create per unit area is very high.

Target Missile

The DRDO uses a modified Prithvi missile as the target for testing the BMD system. DRDO Chief VK Saraswat told the Indian Express in April 2012, "Prithvi missile in a normal course travels only up to

---

2 Shekhar Gupta in the " Walk the Talk" on NDTV 24x7 with DRDO chief V K Saraswat talks about Agni V, India's missile defence system. Accessed through http://m. indianexpress.com/news/we-have-not-understated-the-range-of-agni-v.-we-as-a-nation-dont-have-to-hide-anything-with-respect-to-our-capabilities/943623/

an altitude of 40 Km and covers a range of 350 Km. Whereas I made it go up to a height of 100 Km and come down like a ballistic missile, simulating the terminal velocities of a target of, say 2,000-Km class and then engaging that target with this AAD missile."[3]

Phase - II Interceptor Missiles

The Phase-II missile defence system will be based on the AD-1 and AD-2 interceptor missile that are currently under development. These interceptors would be capable of shooting down missiles that have ranges greater than 5,000 Km, which follow a distinctly different trajectory than a missile with a range of 2,000 Km or less. During their final phase, every ICBM hurtles towards their targets at speeds twice as those of intermediate range missiles. The Phase-II system will match the capability of the Terminal High Altitude Area Defence (THAAD) missiles deployed by the US as part of their indigenous missile shield. THAAD missiles can intercept ballistic missiles up to 200 Km away and track targets at ranges as far as 1,000 Km.

Phase II Radar

The radar system for the Phase-II system will be co-developed within the country with help of Israeli defence industries. Briefing the media about the Phase-II radar, DRDO Chief VK Saraswat told the press on 15 May 2011, "Unlike the Phase-I Swordfish radar developed by India in partnership with Israel, the radar to support Phase-II interception will have 80% indigenous components. Only some of the equipment's and consultancy would be provided by Israel." [4]

Floating Test Range for Phase-II System

A floating test range is being developed as a part of the Phase-II system. During the testing of the Phase-II system, target missiles will be launched from specially constructed ships. Scientists have already started designing the ship and associated systems such as

---

3 Ibid

4 "India developing interceptor missile with 5,000 Km range" by PTI, New Delhi, May 15, 2011 accessed through http://www.hindustantimes.com/India-news/NewDelhi/ India-developing-interceptor-missile-with-5-000-Km-range/Article1-697777.aspx

radar, mission control centre, launch control centre, communication network and many other equipment needed for the trial of phase-II systems.

History of BMD Tests

A total of eight test of the BMD system have been carried out by India, both using the PAD exo-atmospheric interceptor and the AAD endo-atmospheric interceptor. The chronological details are given below:

First Test. The first BMD test was carried out on 06 Mar 2006. During this test a PAD missile successfully intercepted a modified Dhanush surface-to-surface missile fired from INS Rajput (anchored inside the Bay of Bengal) towards Wheeler Island, simulating a target "enemy" missile with a range of 1,500 Km.

Second Test. The second test was carried out on 27 Nov 2006. During this test a PAD missile intercepted a Prithvi ballistic missile at 48 Km altitude.

Third Test. The third test was carried out December 2007. During this test an Advanced Air Defence (AAD) missile intercepted a target missile at an altitude of 15 Kms. The interceptor missile used a 'gimballed directional warhead' or a warhead only one side of which explodes close to an incoming ballistic missile, shattering it. The AAD interceptor has so far been successfully tested up to an altitude of 15 Kms. The interceptor will eventually engage an incoming target missile at 30 Km to validate the efficacy of the missile in its entire endo-atmospheric envelope.

Fourth Test. A test of the AAD missile on 15 Mar 2010 was aborted after the modified Prithvi (Dhanush) missile launched to simulate the target deviated from its flight path. During this test, a Dhanush missile launched from a naval ship was guided along a trajectory similar to that of 1500 Km range Ghauri missile in its terminal phase zeroing in on the Wheeler Island, off Damra village on the Orissa coast. A PAD interceptor launched from Wheeler Island was to intercept the "enemy" missile with a hit to kill at 70-80 Km.

After an in-depth analysis, DRDO clarified that the target missile took off in normal way. However at T+20 sec (approx) the target deviated due to some on-board system malfunction and could not maintain the intended trajectory, thus failing to attain the desired altitude profile. The Mission Control Centre computer found that the interception is not warranted as the deviated target did not present the incoming missile threat scenario and accordingly the system intelligently did not allow take-off of the interceptor missile for engaging the target.

Fifth Test. A test of the AAD interceptor missile was conducted on 26 Jul 2010. This test was partially successful as the missile failed to score a direct hit. A modified surface-to-surface Prithvi missile was launched from a mobile launcher from launch complex-3 of ITR at Chandipur-on-sea. The interceptor AAD missile, positioned at Wheeler Island, about 70 Km across sea from Chandipur, engaged the target missile at an altitude of 15 Km. The warhead exploded within a few metres of the target missile, releasing multiple bullet-like particles which hit and destroyed the target missile 26 seconds after its launch. The significant aspect of this BMD test was that AAD missile for the first time used 'P-charge directional warhead.' This was the fourth consecutive Interceptor Missile test in Endo atmospheric regime at 15 Km altitude.

Sixth Test. An AAD interceptor missile armed with a P-charge directional warhead was successfully tested on 06 Mar 2011. During this test a Target Missile mimicking an enemy Ballistic Missile with a 600-Km range was launched from Launch Complex –III, ITR, Chandipur. The target missile climbed to an altitude of 120 Km and began its downward trajectory. The missile tracking network consisting of long range and multi-function radars and other range sensors positioned at different locations detected and identified the incoming missile threat. The radars tracking the Ballistic Missile constructed the trajectory of the missile and provided the data for continuous complex computations, which were done in real time by ground guidance computer to launch the interceptor missile at an exact time.

The fully automatic launch computer launched the interceptor and the on-board INS (Inertial Navigation System) and ground based radars guided the interceptor to the target (incoming Ballistic Missile).The Interceptor missile intercepted the Ballistic Missile at an altitude of 16 Km and blasted it into pieces. It was claimed by DRDO to be a text book launch and all the events and mission sequence took place as expected.

Seventh Test. The endo-atmospheric (AAD) missile interceptor was successfully tested on 10 Feb 2012. During the test, the AAD-05 missile destroyed a modified Prithvi missile simulating an enemy ballistic missile at a height of 15 Kms. The Prithvi ballistic missile was launched from ITR Chandipur. It was initially picked up and tracked by the LRTR near Puri and thereafter by the Multi-Functional Radar located near the seaport town of Paradip. Based on the tracking information from the radars, a guidance computer continuously computed the trajectory of the incoming ballistic missile and launched AAD-05 interceptor missile from Wheeler Island at a precisely calculated time.

The guidance computer command guided the AAD-05 till the terminal phase of the interception, when a Radio Frequency seeker on the interceptor obtained a lock on the target, enabling the interceptor to hit the target missile directly and destroy it. The warhead on the interceptor also exploded and destroyed the target missile. The Prithvi target missile, mimicking the trajectory of an enemy missile with a range of 2,000-3,000 Km, climbed to a height of 100 Km before descending towards its assigned target. Radar and electro optic tracking systems tracked the missile and also recorded the fragments of target missile falling into the Bay of Bengal. The test was the first in which the interceptor hit the incoming ballistic missile directly and destroyed it. The mission was carried out in the final deliverable user configuration mode.

Eighth Test. DRDO successfully demonstrated the ability of the BMD system to engage multiple targets on 23 Nov 2012. During the test, the BMD successfully intercepted a simulated electronic ballistic missile with a range of 1,500 Km to 2,000 Km at an altitude of 120 Km using a simulated interceptor missile. It also scored an endo-atmospheric hit to kill against a shorter range ballistic missile (a modified Prithvi launched from Chandipur) at an altitude of 15 Km over the Bay of Bengal. The attacker Prithvi missile flew the trajectory of a missile with a range of 600 Km to 1,000 Km. An Advanced Air Defence (AAD) missile launched from Wheeler Island successfully intercepted the 'hostile' Prithvi missile and destroyed it at an altitude of 15 Km.

During this test, two radars processed the simulated and real missiles and assigned launchers to engage them. No interceptor was launched against the electronic missile, but the interception sequence was simulated to T-0 second. Consequent to the test, Dr Avinash Chander of DRDO while briefing the media stated that the missile trial on 23 Nov 2013 was aimed at 'a deployable configuration' to intercept multiple incoming missile threat against India. The system is capable to simultaneously track multiple incoming missiles, process the signals, identify which is a threat and assign the specific launcher-missile that is best suited to intercept them. So far all our interceptor flight-trials have been 'one missile against one target', but this test gave the DRDO team a lot of confidence to simultaneously handle multiple targets.

The outstanding features of the test as told by a DRDO scientist to the press was that teams from five centers (Launch Complex-III at Chandipur; LCC at Wheeler Island, Mission Control Centre, Hyderabad; Long Range Tracking Radar at Konark and Multi-Functional Tracking Radar at Paradip) participated in the successful mission on 23 Nov2012. The SA to the Defence Minister, Dr VK Saraswat told 'The Hindu' newspaper that "The test had demonstrated the maturity of all the BMD technologies, including the directional warhead, radio-frequency seeker as also various networks".[5]

Future Up-gradations for Indian BMD

The Indian defence scientists are working on a laser-based weapon system as part of its BMD to intercept and destroy missiles soon after they are launched towards the country. Dr V K Saraswat of DRDO has explained about the project which is ideal to destroy a ballistic missile carrying nuclear or conventional warhead in its boost phase. As per him it will take another 10–15 years for the

---

5 "India Proves Capability of Missile Defence Systems" by Y Mallikarjun and T S Subramanian in *The Hindu* on 23 Nov 2012. Accessed through www.thehindu.com/news/national/india-proves-capability-of-missile-defence-system/article 4126430.ece

DRDO to make it usable on the ground.

The Laser and Science Technology Centre (LASTEC) is reported to be developing lasers to take out enemy missiles during their boost phase, a stage during which they are most vulnerable. DRDO's Dr V K Saraswat told Press Trust of India in January 2009, "It's easier to kill a missile in boost phase as it has not gained much speed and is easier to target. It cannot deploy any countermeasures and it is most vulnerable at that time".

Appreciating the need of a space based platform for surveillance over ballistic missiles, ISRO is also working on a space-based surveillance system that will help the DRDO in phase-II.[6] Though as of now AWACS airborne radars mounted atop IL-76 aircraft can track missiles above 2,000 km, in order to track missiles with greater range of 6,000 km, the interceptors will take help of radars mounted on satellites. The concept has already been validated by the US 'Space Based Surveillance System' (SBSS) satellites developed by Boeing. The SBSS system will revolutionize the nation's Space Situational Awareness by detecting and monitoring incoming missiles from much more distance and with better accuracy. The US SBSS satellite features a visible sensor mounted on an agile, two-axis gimbal, which allows ground controllers to quickly move the camera between targets without having to expend the time and fuel to reposition the entire satellite. This agile sensor mount enables SBSS to find and track objects in space, even new spacecraft or missile launches and its track.[7]

Indian Strategic perspective of Space Warfare: The Developments

India is developing capabilities and strategies to prepare itself for space warfare in the future. Towards this, a structured approach has been adopted by developing niche space technologies and conducting trials necessary for validation of the same. The main course of action towards fighting a 'space war' in the Indian scenario would be any

6 "India to Deploy Space Based Surveillance Systems for BMD Shield" News accessed through news.xinhua.com/english/2009-03/10/content_10982893.htm

7 Data on Space Based Surveillance System accessed through http://www.boeing.com/boeing/defence-space/space/satellite/sbss.page

or all of the following:

- Defend its satellites against disabling attempts through kill vehicles, laser or other beam weapons.

- Destroy enemy satellites in the orbit.

- Launch mission specific micro satellites on demand.

Hence, in order to secure the vital and sensitive space assets, India needs to work on the above three projects by liberally allocating requisite funds, scientific manpower and the required facilities for time-bound fructification of the required technologies and testing validation of the same towards creation of the desired and a reliable system. These projects are briefly discussed below:

Anti-Satellite Capability

After the successful maiden test launch of the Agni-V missile on 20 Apr 2012, DRDO Chief Dr VK Saraswat reiterated during a press conference in New Delhi that the Agni missiles have anti-satellite capabilities. He said Agni-V can provide the necessary velocity and range to reach the needed altitudes. DRDO also had the guidance capability to direct the warhead towards the intended satellite in space and eventually destroy the satellite using a 'kill vehicle' or just disrupt the satellite's functioning.

Satellite Defence Capability

India is developing capabilities to defend its satellites against attempts to disable them using kill vehicles, laser or other electronic weapons. It has also developed the co-orbital manovering capabilities, either to use its satellites as co-orbital ASAT weapon or to take evasive defensive action by changing the orbit of the satellite, thereby evading ASAT attack. Another defensive action which can be taken to protect the satellites from adversarial attack is 'space mining'. India has already developed and demonstrated its multi-satellite launch capability. It has on 28 Apr 2008 successfully launched 10 satellites in a single mission, boosting its capabilities in space. [8] The technology was re-validated again on 25 Feb 2013

---

8   BBC News, 28 April 2008, "India in multi-satellite launch", accessed through http://

when in a multiple launch mission, a Polar Satellite Launch Vehicle (PSLV-C20) put India-French satellite SARAL and six others into their precise orbits after its successful launch from the Sriharikota spaceport.[9] This multiple launch capability could be used for 'space mining' to protect its space assets from ASAT attack.

Satellite Launch "On Demand"

On 31 Mar 2012, DRDO for the first time announced at a press conference during DefExpo 2012 that DRDO is building up capabilities to launch small satellites on demand to support the armed forces. The capability will provide communication, navigation and guidance support to the armed forces during crises. The launch capability will be based on Agni-IV and Agni-V missiles, which will launch mini- and micro- satellites within few hours of demand.

During a press conference in New Delhi on 20 Apr 2012, a day after the successful maiden launch of the Agni-V missile, DRDO Chief VK Saraswat reiterated that his organization's intends to develop 'on-demand small satellite launch capability' using Agni missiles. The capability would help India to place mini- and micro-satellites in orbit as replacements for any critical navigation or communication satellite disabled by the enemy. The micro-satellites would have a short life span of between 6 months to a year life.

Conclusion

With the growing economic and military dependency on space, no space faring nation can afford to leave its space assets neither undefended nor leave itself open to attacks from space. In this regard an analogy has been drawn between sea and space. Just as nations developed navies to protect their investments transiting through sea, similarly it is imperative to develop a robust space force equipped with space weapons, to protect national interests in space. This poses a great challenge to the notion of 'space as global common', which accorded 'freedom in space' to all and professed 'space for peaceful purposes'. Even though there is an international body for laws

---

news.bbc.co.uk/2/hi/south_asia/7370391.stm

9   TS Subramanian in 'The Hindu' 25 Feb 2013, "PSLV-C20 puts SARAL, 6 other satellites in precise orbits" accessed through http://www.thehindu.com/sci-tech/technology/pslvc20-puts-saral-6-other-satellites-in-precise-orbits/article4452165.ece

governing 'freedom of the seas' UNSLOC, yet countries knowingly defy these laws during war or at times as per their requirement. Thus, it is just a matter of time that space will be weaponised by states with a sole aim of occupying the 'ultimate high ground' for pursuance of national ambitions.

Gray and Sheldon summarized the situation very aptly in their article "Space Power and the Revolution in Military Affairs: A Glass Half Full," when they wrote:

"Space control is not an avoidable issue. It is not an optional extra. If the U.S. armed forces cannot secure and maintain space control then they will be unable to exploit space reliably, or reliably deny such exploitation to others. The U.S. ability to prevail in conflict would be severely harmed as a consequence. If you fail to achieve a healthy measure of space control in the larger of the possible wars of the 21st century, you will lose".[10]

Thus in future wars, the 'Battle for Space' may be analogous to the 'Battle of the Sea'. Hence, developing and deploying space weapons to safeguard national interest would be the most logical conclusion for India. The bottom line is that space control is not optional but inevitable. Let's not forget that 'A nation can get peace, only if it can defend peace'. Thus it would be prudent for India to develop the required ASAT and BMD technologies, so as to demonstrate these capabilities and to deter any adversary from undertaking any mis-adventure against the country.

The Chinese direct-ascent ASAT test raises difficult questions about China's intentions towards its potential adversaries, the US and India. Since the US has a range of potential responses to Chinese ASAT capabilities, it can defend its national interest. Unfortunately, there are no options available to India to defend its space assets as of now, simply because no single option is simple, cheap, or likely to be wholly effective. Therefore Indian policymakers should consider limiting Chinese deployment of ASAT capabilities and the technical/operational measures that would mitigate the impact of Chinese ASAT on the Indian space assets which can hamper the economic

---

10 Colin S. Gray and John B Sheldon, "Space power and the revolution in military affairs: A glass half full?" in Airpower Journal; vol 13 no 3; Fall 1999; p23-38

and military operations.

Today India stands in league with most advance space faring nations of the world. It is a matter of pride for the country that its launch capability finds trust in international market and countries including US, Canada, France and Israel, are availing Indian satellite launch facilities, which are the cheapest in the world. As regards satellites, India today boasts of having the best and the largest number of remote sensing satellites in orbit. Besides, the country has done wonders in the communication satellites, weather satellites and navigational satellites. The second decade of the 21 st century also saw the launch of spy satellites such as the RISAT – I on an Indian rocket PSLV- C19. But for international sanctions, India would have been in the forefront of space exploitation, perhaps at par with the US, Russia and China.

Notwithstanding these difficulties, India launched a road map for its space activities and finalised a blue print in the form of Vision 2020 for space. This document has outlined many ambitious space projects for ISRO, which when achieved will put India into the elite group of space faring nations. This will also enable India to utilise space in the best possible way for not only trade and commerce, but also for strategic and tactical planning.

In the emerging geo-political and geo-strategic scenarios India cannot afford to escape from developing a credible and a potent ASAT capability, keeping in view the large fleet of satellites it operates today and the belligerent Chinese approach towards it, especially after the ASAT test in 2007. Thus to defend its space assets, India doesn't have a choice, but to eventually work towards a two pronged strategy to counter the threats emanating from the Chinese ASATs, which are to be taken up concurrently

- Militarisation of Space.
- Weaponisation of Space.

The time for this is running out for India as the revised Code of Conduct for Outer Space activities proposed by European Union is coming up for debate in the UN. If this is ratified in the near future, India's options towards militarization or weaponisation will be curtailed by the UN diktat. If India cannot develop and test its ASAT capabilities by then, it will have to face "Space Apartheid" as it experienced in the case of "nuclear apartheid" at the hands of the P-5 nations. Hence, a credible and potent Indian ASAT will be a deterrence for China and will act as an insurance against any strike against Indian space assets. This will ensure security of Indian space assets, which are 'engines for economic growth' and in turn render peace and prosperity, not only for India, but also for the entire Asian continent.

Index

www.ingramcontent.com/pod-product-compliance
Lightning Source LLC
Chambersburg PA
CBHW050528190326
41458CB00045B/6756/J